D0457663

MANAGING AGRICULTURAL SYSTEMS

MANAGING AGRICULTURAL SYSTEMS

G. E. DALTON

Head, Agricultural Economics Division, School of Agriculture,
Aberdeen, UK

APPLIED SCIENCE PUBLISHERS

LONDON and NEW YORK

APPLIED SCIENCE PUBLISHERS LTD
Ripple Road, Barking, Essex, England

Sole Distributor in the USA and Canada
ELSEVIER SCIENCE PUBLISHING CO., INC.,
52 Vanderbilt Avenue, New York, NY10017, USA

British Library Cataloguing in Publication Data

Dalton, G. E.
    Managing agricultural systems.
    1. Farm management
    I. Title
    630′.68   S561

ISBN 0-85334-165-6

WITH 26 TABLES AND 26 ILLUSTRATIONS

© APPLIED SCIENCE PUBLISHERS LTD 1982

Printed in Great Britain by Galliard (Printers) Ltd. Great Yarmouth
(Photoset in Malta by Interprint Limited)

Acknowledgements

I should like to thank three groups of people who have helped me in the writing of this book.

My colleagues and friends in the Agricultural Economics Division of the School of Agriculture, Aberdeen, and my former colleagues in the Departments of Agriculture in Reading University and the University of Ghana—in particular, the discussions I have had with Peter Charlton, Bob Crabtree, Tony Giles, John Marsh, Brian Pack, Blair Rourke and Martin Upton—have been most helpful.

My neighbours' support has been invaluable over the last six years. They have helped to create an environment in which it has been possible for me to learn, often the hard way, a great deal about real agricultural systems. The advice and support of Jimmy Adam, Bruce Jaffrey, Duncan McConnach and John Stephen is appreciated.

My family too, even in my moments of doubt, have always had faith in this project.

Preface

A large proportion of the world's population is directly involved in agriculture, many of them as small businessmen. Each of these businessmen applies his skill and resources day in, day out, to produce food and raw materials. All of them to a greater or lesser extent operate in a changing environment to which they must adjust

Many aspects of the environments of agricultural institutions are influenced by the decisions of other people; in the marketplace, in ancillary industries, in research institutes and places of learning and in Government organisations that create and implement agricultural policy. Improvements in the quality of decisions made by those who manage all the various parts of the agricultural system can have a major impact on the quality of everyone's life. How can such improvements be achieved?

This book is based on the belief that a grasp of concepts or principles can add a great deal to the understanding and management of agricultural systems of all kinds. It therefore attempts to transmit some ideas and observations about management in agriculture. It is not about techniques nor about the particular problems of a specific type of agriculture. Its aim is to promote reflection, thought and discussion by professional people whose decisions influence the performance of agricultural systems. As such, the book attempts to convey only the central concepts of management in as simple a language as possible.

G. E. DALTON

Contents

CHAPTER 1

Introduction

GENERAL MANAGEMENT PROBLEMS

The manager in agriculture contributes either directly or indirectly to the production and distribution of food and raw materials by using resources. This transformation of resources into products which people want and are prepared to pay for is not an automatic process. It takes place because those in agriculture work and worry about the three classical questions posed by the theory of production, namely what to produce, how much to produce and how to produce.

Managers have to decide on the quantities of different goods and services that are produced and traded. They can change their systems of production in order to increase the amounts of more valuable types of output. The requirements of consumers can be met more closely by improvements in the quality of produce and by changes in the timing and location of activities. The development and adoption of new methods can improve the inherent technical efficiency of production and enable more expensive inputs to be replaced by less expensive ones.

The challenge for management is to simultaneously find a combination of enterprises, methods and resources (a system) which can be operated in such a way that the benefits derived exceed the costs if not by as much as possible then certainly by a satisfactory amount. Strategic and tactical decisions alike have to be made within an uncertain environment and their consequences accepted over time.

MANAGEMENT PROCESSES

A generalised description of management processes is presented in Fig. 1.1. It shows the various processes that are involved in making a system work.

Plans to change a system are based on forecasts of the future state of the environment and the system itself. Acceptable plans have to be implemented. The recording of the output from a system provides information about problems which can be corrected either by making new plans or by control processes. Anticipatory control also requires forecasts of future states of the system and any type of control process involves corrective action. All these processes interact and are continuously taking place.

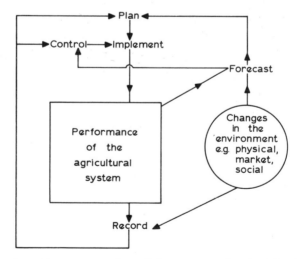

FIG. 1.1. A schematic representation of the processes involved in the management of an agricultural system.

The whole set of management processes are carried out for a purpose. In this particular description the setting of objectives has been included as part of the planning activity since no planning problem exists unless a system is not meeting the objectives set for it. The setting of objectives in fact belongs to a higher dimension in life than any management activity as depicted in Fig. 1.2.

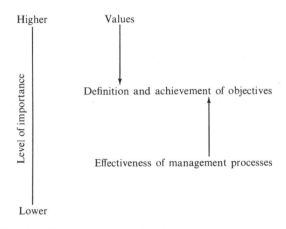

FIG. 1.2. The importance of objectives in management.

OUTLINE OF THE BOOK

The analysis presented in Fig. 1.1 provides the structure of this book. Since the respective processes are carried out in an iterative manner there is no logical order of presentation.

The book begins with an attempt to define a system (Chapter 2). Since systems are mental concepts and their construction depends on the point of view and purposes of the manager, no attempt has been made in the book to classify systems into particular types. This may cause the reader some difficulty since we all construct systems within our own discipline and we tend to differentiate between the general and particular. The systems approach, however, is sufficiently general to apply to the problems of any type of system. Its contribution to management is that it emphasises interactions between the components of systems, it stresses the importance of objectives and it encourages the study of problems as a whole including dynamic and uncertain elements. The chapter on Agricultural Systems ends with a description of modelling procedures which are used in studying agricultural systems.

In Chapter 3 the performance of agricultural systems as affected by changes in the environment is discussed. The environment is defined as factors which influence the system but are outside the manager's control. Important elements in the agricultural environment include the weather, the operation of agricultural markets and the state of domestic and international economic affairs. Finally, there is a brief reference to the changing attitudes of society towards agriculture.

The components of the planning process are described in Chapter 4. The chapter begins by referring to the fact that information which does not change plans, has no value. The importance of objectives in planning is stressed and the difficulties of defining and reconciling competing objectives including time preference is explained. The planning sequence begins by a thorough search for alternative ways of doing things and is followed at both the strategic and tactical level by identifying the best solution proposed. Some techniques which assist the planning process are described including partial budgets, linear programming and simulation.

The principles of control are considered in Chapter 5. Management will have standards for system performance and deviation away from this standard will at some point produce corrective action. Deviation can be controlled in closed loop systems by changes in the output level producing appropriate adjustments to the controlling input levels through feedback mechanisms. In open systems deviation can be controlled by compensating for disturbances or by eliminating them. The principles of building reliable systems from unreliable parts are explained by reference to the duplication of components in sequential processes. The way in which diversification and storage reduce variation is also analysed.

Measuring system performance by the use of records is described in Chapter 6. Information is built up about systems by recording events. Basic records can be checked and built up into more general information including relationships. The construction and use of a profit and loss account and a balance sheet illustrates this process. One use of such information is to compare performance between similar systems. However, comparison between institutions is only valid when homogenous groups can be identified. The problem of comparability also occurs in the application of experimental results and in the gathering of survey data.

Forecasts (Chapter 7) are necessary because actions taken in the present produce outcomes in the future. As a result management can never be sure in advance of success. Information is subject to error, relationships may not be quantified accurately and they may be applied incorrectly. The effect of changes in the environment depends on which events occur, their timing and their impact. There are therefore several assumptions made in deriving forecasts and these can be presented systematically using decision trees. A problem for management is to reduce the variety of forecasts to a manageable number. Possible outcomes can be weighted by estimates of their likelihood or events ruled out which are thought to be unlikely or of little effect. While the future is uncertain, limits on possible future states are restricted by present conditions and known relationships.

The quality of work in agriculture can have a great effect on performance. Ways of creating conditions in which people can work well is the subject of Chapter 8. The subjects covered include work organisation, motivation and ways and means of avoiding inequity both in the marketplace and between small and large producers. The systems concept of breaking down a person's job into its constituent parts in order to identify training needs is described. The capacity of communication channels is explained and the need to organise agriculture so that the social repercussions of independent activities can be controlled is also referred to.

CHAPTER 2

Agricultural Systems

DESCRIBING AGRICULTURAL SYSTEMS

Agriculture is a collection of activities which converts mainly biological inputs into food and raw materials. The activities can be classified into a set of sequential processes such as input manufacturing and distribution, farming, food processing, wholesaling and retailing. Within each of these sectors managerial decisions are formulated and implemented in response to changes in other sectors in the sequence and also to changes which take place in the rest of society. The integration of the activity in all the components of agriculture gives rise to a set of products and services at each stage in a vertical production and marketing chain which are passed on, after modification, until they reach the consumer and disappear.[1]

Another way of looking at agriculture is as the combination of activities classified according to the skills required. They include manufacturers, engineers, salesmen, auctioneers, farmers, hauliers, teachers, researchers, advisers, administrators, contractors, politicians, officials, veterinarians, tradesmen and accountants. The list is not exhaustive but it does illustrate that one of the major functions of management is the integration of skills for a common purpose.

The view of agriculture varies according to perspective. The problems facing a Minister of Agriculture are different from those of a farmer or a large machinery manufacturer. The Minister's influence on the whole system will be broad and general. The farmer's much less so. Indeed it may be so slight as to be safely ignored unless his state is common to many farmers. The large machinery manufacturer can have a direct influence on the design, safety and operational characteristics of the machines made by him and indirectly on machines produced by other firms and ultimately on the resources used and the type of production that takes place.

Systems are mental concepts devised for specific purposes. The components within them depend on what can be controlled and the objectives of the particular manager. Consider a system called a cow. It will be described in different ways by a stock person, a veterinarian, a nutritionist, a rancher and a tourist. Individual cows may be too detailed components for the purposes of defining national agricultural systems and too general in descriptions of biochemical processes. There are differences in the levels of systems ranging from the very general to the highly particular as illustrated in Fig. 2.1.

The search for causes and explanations tends to result in the construction of more detailed systems which makes it difficult to be sure that analysis is taking place in the right area of knowledge.

Systems are simplified in two main ways. Firstly, the quantities or components of systems are classed as either environmental or endogenous. Environmental quantities unlike endogenous ones are given and are not controllable by the manager. Secondly, the number of components for consideration is reduced by classifying them into homogenous types according to definitions. For example, the definition of a farm may exclude gardens, small holdings and intensive poultry units but treat in the same way production units with widely different characteristics of size, ownership, inputs used and outputs produced.

The endogenous components of systems are further classified into inputs and outputs. The same real component can have a different designation according to the system under consideration

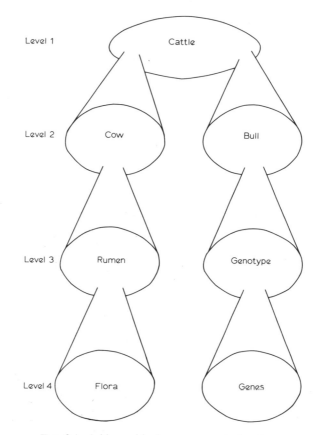

FIG. 2.1. A hierarchical arrangement of systems.

since the output from one type of firm for example can be the input in another type of firm, e.g. cattle and grain. Inputs may be also designated as either being stock (or fixed) resources or flow (or variable) resources.

Fixed resources are not used up within a single production cycle. They tend to be made available in discrete units and last a long time. They are not specific to the production of one type of product. Examples include buildings, equipment, machines, fences, roads and land improvement. Variable resources such as fertiliser, feed, chemicals, raw materials of various kinds, are more or less used up within a single production cycle and are specific to particular forms of output. They are processed through the fixed resources. The distinction, however, between fixed and variable resources is not strictly true. Some part of a building or machine is worn away each time it is used. Fertilisers and feed, for example, are not completely used up by a single crop or an animal and a significant amount of nutrients are re-cycled either as crop residues or animal manure.

Managers change a system's output by choosing one or more activities and by manipulating the attributes of the components required for these activities. The amounts, cost, priority, timing, type, speed, colour, genotype, combination, frequency and duration of components are all examples of the type of variation management can exercise. The effect of changes in component characteristics on system performance is achieved through relationships. It is not necessary to understand the reasons for relationships provided management knows from experience that events are empirically linked to other events. Less error is likely in management decisions where causality is understood. In either case it is important for managers to know that relationships exist. Many problems such as pollution and poverty are compounded by ignorance. The degree by which system performance can be changed is restricted by the form of relationships be they linear or curvilinear and by the size and sign of their coefficients.

The ability to change system performance is also governed by the initial state of a system. It is hard, for example, for a firm in financial difficulties to acquire resources for investment. Relationships within agriculture tend to feed on themselves so that institutions are either in a vicious or a virtuous circle. Fertile pasture allows farmers to stock more intensively so that more manure is produced which ensures, in turn, higher fertility. A degenerative cycle of low fertility, low stocking and inadequate manure can just as easily be produced. The momentum of a system thus determines future possibilities.

Constraints of various kinds also limit the manager's freedom to make changes. These may consist of simple identities such as the demand for resources cannot exceed the supply of resources. Constraints are imposed by the legal system, moral values and by the technical relationships that exist between activities as for example in rotational practices. Capacity limits will also exist although these may be stretched within given bands by altering operating procedures, for example by working faster or ensuring greater throughput and turnover.

Finally, systems can only be defined in relation to the objectives of management. They determine the level of abstraction chosen, the relevant components and their definitions and operating characteristics, while the output from constrained relationships cannot be set without applying some kind of choice criteria.

A generalised summary of the ingredients of a system is shown in Table 2.1.

TABLE 2.1
The Ingredients of a System

1. A set of assumptions
2. A set of components or objects: Inputs and outputs
3. A set of definitions
4. A set of theorems and identities
5. A set of relationships and constraints
6. A set of attributes or characteristics
7. A set of activities or processes
8. A set of objectives
9. An environment or things outside the system

CHARACTERISTICS OF AGRICULTURAL SYSTEMS

Complexity

Agricultural systems are complex. Many components interact with each other in various activities and each component can have several important attributes. It is therefore feasible for a system to be in one of many different states at any given time.

Imagine a farmer with two fields, A and B, to harvest. The fields over time can take on four different states. Both fields can be unharvested, $A_0 B_0$, or both fields can be harvested, $A_1 B_1$, or only one field may be harvested giving two possible states, $A_1 B_0$ and $A_0 B_1$. If we now add a third field into the system there are now six different orders of harvesting fields, namely, ABC, ACB, BCA, BAC, CAB and CBA. For each order there are eight different possible states as shown for order ABC below:

$$A_0 \; B_0 \; C_0 \qquad A_1 \; B_0 \; C_0$$
$$A_0 \; B_0 \; C_1 \qquad A_1 \; B_0 \; C_1$$
$$A_0 \; B_1 \; C_0 \qquad A_1 \; B_1 \; C_0$$
$$A_0 \; B_1 \; C_1 \qquad A_1 \; B_1 \; C_1$$

(The variety,[2] or the total number of states of a system is given by the formula variety $= x^n$, where n is the number of components and x is the number of possible states of each component. In the example, 3 fields have 2 states, i.e. variety $= 2^3$.)

The onlooker monitoring the progress of this particular farmer's harvest could conceivably see one of $6 \times 8 = 48$ possible states. The number of states therefore increases exponentially as more fields are added and the greater the number of field conditions.

This extremely simple example illustrates how the complexity of a system can be measured by the total number of different arrangements. It shows the very large number of ways that potentially exist for organising agricultural systems. Many of the actual arrangements may not be important or may not be feasible. Fields do not ripen all at once. Indeed positive steps are taken to make sure that they don't by altering variety of seed and sowing dates so that complexity is reduced. Nevertheless, in the above example, unless some rules or operating procedures can be devised or the problem can be avoided in some way, one can imagine the almost permanent state of indecision in a combine driver's mind about which field to cut next.

Coping with Variety

One function of management is to cope with system variety. This can be achieved either by reducing variety or designing management procedures which have equal and opposite variety.

Fixing the values of some attributes can greatly simplify the number of possible system states. This is why the initial state of a system is so influential in determining future situations. In the example of the ordering of fields to be harvested, if field A is cut, then the possible orders which remain are reduced to ABC and ACB and the states to four for each order, since A has occurred. Thus 48 possible states are reduced to eight once one field is cut.

The number of states for consideration can be cut down by focusing only on those components which have a major influence on performance. Another way is to concentrate on those components which change and assume that the rest of the system is unchanged.

Understanding variety is also helpful to management. A system is not just a collection of components; it works. This means that the parts of the system are arranged in a particular order and that one system state is not followed by any one of all possible remaining states but by a particular set. Changes in a system are conditional on some other state having occurred previously or taking place at the same time. Association of particular system states with performance helps management to recognise or anticipate problems. Rather than observing all system states all the time management can watch out for those system states which are connected with unusual or preferred performance levels.

One of the remarkable features of agricultural systems is the balance that is achieved between their different parts. This balance is even more remarkable given the very large numbers of people involved and the wide geographical distances that exist between producers and eventual consumers. There is also a wide distance between producers and consumers in an information sense since with increasing specialisation very little is known by consumers about the way in which food is produced either in the processing factory down the road or on farms and in manufacturing plants in other continents. The features of climate and soil largely determine where different types of animals and crops are grown and kept. Many of the products are perishable and costly to store, process and transport and yet for the most part the effective wishes of consumers influence production and marketing decisions quickly and directly both through the price mechanism and other forms of communication. The infinite complexity as measured by the possible number of different food mixes in terms of type, quality, timing and continuity, eaten by hosts of consumers, is matched by the myriads of production decisions made by all those further back in the production marketing chain.

Order reigns in spite of the complexity by the balance of variety between producers and consumers. The requirements of consumers can only be met by the ability of producers to meet them, so that a state of equilibrium is achieved.

Complexity is a necessary attribute of systems. The successful management of variety will enable the objectives of many different people to be met. Management has available a wide array of system designs and methods of operation. The job of management is to choose and run the most appropriate system.

Constraints on System Choice
The inherent characteristics of agricultural inputs influence the way they can be used. The fact that they complement each other means that it is quite possible that production may be limited by the availability of one particular input. The response in output to a change in the level of a single input will depend on the level and variation in the level of all other inputs. In practice it is not always possible to vary one input at a time since they are available in the form of joint packages. Thus, different nutrients in various proportions are present in manure and animal feeds. Likewise inputs such as machinery and equipment often incorporate new technological ideas. People who are employed because of their ability to do a particular job often have other desirable and undesirable attributes. Nor is it possible to produce precisely what is required. For example, mutton cannot be produced without wool, beef without hides or grain without straw. More generally, high quality produce cannot be obtained without at least some low quality produce.

The complementarity between the elements combined in joint inputs and products is also a characteristic of biological relationships where the output from one activity enhances the output from another. The best example of this is in the quality of leguminous plants to fix atmospheric nitrogen. It is an advantage to grow other crops with legumes either in the same mixture or at a subsequent stage in the rotation. The use of by-products is another case of the importance of fitting activities together to avoid waste. In the processing sector this can often depend on having a sufficient amount of by-products available in one location in order to justify the installation of specialised plant, for example, the by-products of the slaughter industry. The same kinds of arguments also apply to grading operations. Selecting output for quality in order to satisfy a discriminating market means that the overall returns depend not only on the price of the high valued product but also on the value of the second rate. The lower valued product may have a different market or may have to be used in a different way altogether. Another complementary relationship occurs with flows of cash into and out of a business. Returns from one activity will finance the inputs of another in an ongoing business. Firms starting up operations will not have this in-built stability so that their cash requirements and vulnerability are all the greater.

The relationship between inputs and outputs within any productive process is normally governed by the principle of diminishing returns. This states that with other factors held at a constant level the response to extra units of input will eventually decline and indeed may become negative. In practice in many situations the onset of diminishing returns can be fairly abrupt and the process can be represented by a bent stick type of function as shown in Fig. 2.2. Once a limit is approached the rate of output change can be zero or very small. Eventually in any situation a point will be reached when output falls.

One of the problems of management is that whereas response functions can be defined under strictly controlled conditions, in practice the assumptions of conditionality are broken as all input levels can change. Some evidence shows that a great deal of flexibility exists in production systems. The same level of profit can be made in many different ways as shown by the continued existence of firms in the same line of business with markedly different systems of production. It is extremely dangerous to pass judgements on efficiency on limited or partial data, e.g. on

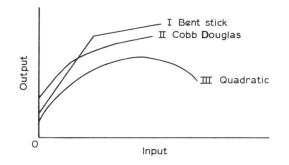

FIG. 2.2. Simple forms of relationships.

yield levels or revenue less directly attributable costs (Gross Margins).

Scale

Allowing all input levels to increase simultaneously results in the growth of an enterprise. Where economies of scale exist the cost of production is on average lower than in smaller units. One of the reasons for this can be demonstrated for buildings. Construction costs are lower per unit of capacity the greater the size of the building since the common costs of walls and roof rise at a linear rate with scale but capacity increases at an exponential rate.

There is in practice some confusion between economies of scale and economies of utilisation. For an input like a machine the cost per unit of output as it is used more fully will decline, initially quite sharply and eventually slowing down as shown in Fig. 2.3. To achieve such economies a minimum size of production unit is required. (In Fig. 2.3 costs fall sharply up to 100 units.) This may not be feasible in small firms given the difficulties of acquiring sufficient resources. Nevertheless it is important in many parts of agriculture to exploit capacity to the full by ensuring high levels of throughput for such facilities as slaughterhouses, packing plants and retail outlets. If capital is a restricting resource then achieving high levels of throughput is the main way in which high returns can be acquired.

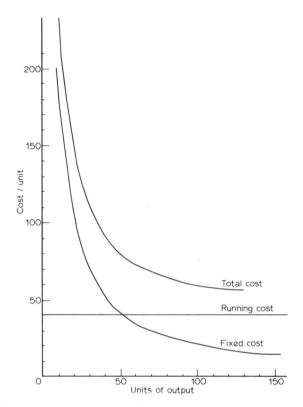

F IG. 2.3. The relationship between utilisation and unit costs for a machine. Fixed
costs = £2000; running costs = £40/unit.

Time

Throughput measured as output per unit of time is important in
most stages of agricultural systems. Sales per unit of shelf space in
a supermarket is a common management criterion. The number of
animals put through a slaughterhouse or a fattening unit or the
number of tractors made per day are all different measures of the
same concept within different stages of the agricultural process.
Production in agriculture is seasonally based and this restricts the
extent to which machines and other facilities can be utilised.

During the season inputs must be available including finance,
labour and direct inputs and these can be acquired either by
paying high prices or by storing them until they are required. The
matching of seasonal production to the demand for products is

also carried out through storage operations. Seasonality can result in output being restricted simply because of the availability of resources at critical times of the year. For example, in shifting cultivation systems, the amount of land that can be cleared in the dry season determines the area planted. This area may not mature if it cannot be weeded later on in the season. In locations where two growing seasons occur within the year, the time between harvesting the first crop and planting the second is often short. This also applies in colder climates when crops are planted in late autumn.

Time is also needed for crops and animals to grow which means that supplies cannot suddenly be increased. Some supplies have to be kept back for seed or for breeding purposes so that the first response to extra demand is a fall in supplies. The age structure of crops and animals is also important since productivity, as measured by yields, first of all increases with maturity and then declines.

The timing of events is also important in view of the fact that there is a wide diversity of specialist activities within agriculture, operating in sequence. For example, any interruption in the flow of soya meal from the major exporter, the USA, to the world's poultry producers would be highly disruptive. This is illustrated for an actual case on an African poultry unit in Fig. 2.4. Pig

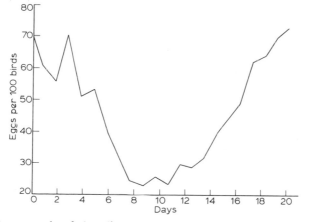

FIG. 2.4. An example of the effect of an interruption in feed supply on layer production.

producers also depend on a supply of highly bred replacement stock, specialist equipment and management, power supplies on a continuous basis, transport, information, drugs and processing facilities. A great inter-dependence exists at all stages in the total system.

Stability

Maintaining stability in inter-dependent systems is all the more difficult given the wide variation that exists in weather, prices, technical performance and characteristics of populations within the agricultural industry. Frequency distributions often approximate to the normal curve as for example for yields, growth rates and grades but skewed distributions are also common. The distribution of farm size, for example, is highly skewed, a few farmers being responsible for a more than proportionate share of agricultural output and many small farmers making a much less than proportionate share to total output. At a more detailed level, small machinery breakdowns tend to occur much more frequently than large ones.

Variation in quality such as size, appearance, genetic potential, growth rates and timing enables agricultural marketing to be discriminatory according to the needs of various markets. This can be done by grading, selection and even through planning where the nature of variation is well understood. Such variation can give rise to capacity problems as will be explained in later chapters. In pig production, for example, the pressure on fattening accommodation will depend on the number of pigs that farrow per unit time, the size of litters, and the growth rate of both weaners and fatteners. A sequence of large litters or a batch of slow growing pigs can throw the whole system out of balance. This is more likely to occur the greater the utilisation of facilities as less flexibility is available.

Technological Change

The confidence and capabilities of system operators can be easily overlooked. New practices add to the already considerable uncertainties of agriculture and even when change is welcomed it takes time to experience and practise new technology. There are several important features of technological change in agriculture. Firstly, in recent decades the pace of change has been rapid in most countries of the world. It has been the main determinant of increased agricultural production. The main feature of new technology in agriculture is that it increases output. If this extra output is produced in a market with a static demand then less resources are needed within agriculture so that structural adjustment problems, especially for labour, are created. The driving force for new technology in agriculture is innovators' profits, that is, those profits which accrue to producers for being first in adopting methods which lower their costs while market prices still reflect the higher average costs of most other producers. The main long run benefits of technological progress in agriculture accrue to consumers. These may be in distant parts of the world but even so, out of the need to encourage exports and to lower prices for domestic consumers or to release resources from agriculture for use elsewhere in the economy, governments generally encourage research and the adoption of new technology. Governments very often need to take part in such activity because agriculture is organised in small units which are insufficiently large to undertake research on anything other than a collective basis. This does not of course exclude large firms who service agriculture from conducting commercial research. Their incentive is to develop new products so as to share in innovators' profits.

The Market
The overriding influence on the development of agriculture is the state of the market. The market or its substitute has to work. Inputs must be available and outlets must exist for the products of agriculture. In many parts of the world this is not the case. Large publicly financed projects such as irrigation may not be available. Extension work may be of indifferent quality or research work inadequate or irrelevant. Important supplies such as spare parts, fertilisers, chemicals and feed may be the subject of foreign exchange regulations or import licences. Transport may not be available because of inadequate roads and shortages of vehicles.

Credit may only be available on extortionate terms because of the lack of competition between credit agencies and high risks of default.

Prices within the agricultural market must also be sufficiently attractive to provide incentives for development. Most investment funds in agriculture are generated out of income and unless incomes are adequate to meet consumption needs development will not take place. The ratio of input prices to output prices is all important especially as the level and trends of prices also influence confidence. Confidence among the managers of agricultural systems combined with attractive prices can result in rapid increases in output. In many countries markets are tightly controlled either through quotas, price subsidies or taxes or in centrally-planned economies through more direct forms of regulation so that the actions and intentions of governments are an important influence on the progress of agriculture.

MODELS—A WAY OF THINKING ABOUT SYSTEMS

Models are simplified representations of reality. They vary according to their purpose and to the degree of abstraction. A simple example of a model is a map where streets, buildings and points of interest are marked by lines, symbols and colours. Maps save a lot of time and enable people to move around more easily and to plan changes in the arrangement of facilities. There are many different kinds of maps. There are maps of the world, of countries, of regions, of towns and parts of towns. The trick is to choose the right map for the right purpose. A map of the world is of very little use to a person lost within a town and a map of a town is of no value to a person lost in the world. An array of different models is available ranked according to their degree of abstraction. At the most general level many elements are coalesced or mapped on to one part of the model. At the most detailed level there is a one to one correspondence of elements in reality and elements in the model. It is useful when analysing problems to start with a fairly general model so as to identify the areas in which more information is required, otherwise there is a danger of getting lost in detail.

Types of Model
There are three main types of model, namely, iconic, analogue and symbolic. Iconic models usually preserve the same appearance as the real system except for their scale. They are usually physical models. Examples include the plots in experiments which represent fields, models of machines and equipment and models of animals which are used to study heat loss and requirements under different climatic regimes. They tend to be inflexible so that experimentation is time-consuming. They are also expensive but are very good representations of reality.

Analogous models are more abstract than physical models since the objects in the model are not made to look like the objects being represented. The effectiveness of analogous models depends on how well the analogy can be understood and how well the behaviour of the real system corresponds to that of the analogue. A good example would be that the administration of a large organisation is analogous to the operation of the central nervous system in an animal. An organisation therefore requires sensory mechanisms (eyes) in order to gain intelligence about the outside world. It needs to be able to transmit information within itself (nerves) and to sort, memorise and retrieve information (brain) as required by the decision making parts of the organisation (mind).

The objects and interactions in symbolic models as the name implies are represented by symbols. On a map lines and squares represent streets and buildings. The usual symbols are mathematical ones using algebraic symbols and numbers. Accounts can also be classed as symbolic models where flows of goods and services and stocks and assets along with liabilities are handled in a specifically defined way as numbers. Symbolic models are general, flexible and their construction forces management to be systematic and rigorous in the definition of their problems. Once built they can be used for several purposes including planning, control and forecasting (for examples see Dalton[3]).

Features of Models

Models are often classified according to whether they include time, risk, linear and non-linear relationships within their structure. Some models assume that changes take place in a continuous way while others allow discrete or abrupt alterations in the level of variables. The behaviour of model output can demonstrate both stability and instability as measured by returning to the original system state after the effect of a disturbance. Steady-state behaviour is where a model's output is repetitive with time whereas transient behaviour describes those situations where the character of output variables changes with time, e.g. in a model of growth.

Models which are easy to solve are normally static, linear, continuous, deterministic and stationary. Thus in a linear model, situations are purely additive. The amount of calculation is much reduced if the rules of calculus can be used without the complications of random variables and time. Unfortunately, the assumptions necessary to build such simple models can be so unrealistic as to reduce the value of the modelling process. Real life is discrete, transient, stochastic, non-linear and dynamic. Detailed models of such systems are large, require a lot of data, are expensive to run and difficult to solve. However, the criticism of models because they are unrealistic ignores the fact that the reason for modelling is to simplify reality while retaining the important characteristics of a system. In this way variety is reduced to a manageable magnitude. The simpler a model is, the better. Investigations are carried out with models as shown in Fig. 2.5 rather than with real systems. It is usually less expensive, quicker, less dangerous and necessary if the system does not actually exist.

Experiments with Models

Experiments with models can be done in four different ways, namely, as an operational exercise, as a game, through simulation and finally analytically. The features of these different approaches are described in terms of speed of analysis, degree of abstraction and cost, in Fig. 2.6.

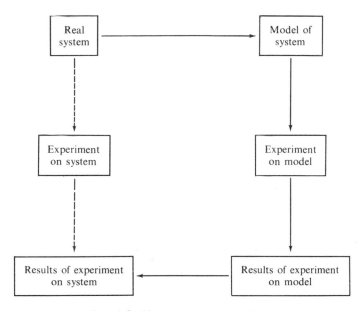

FIG. 2.5. The process of modelling.

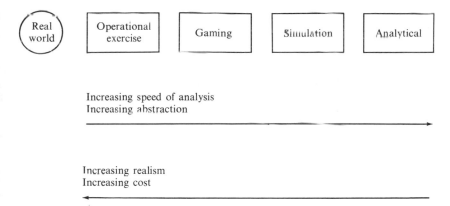

FIG. 2.6. A spectrum of models.

An operational exercise is expensive but is employed by management when pilot plants are constructed or where new developments are first of all initiated on a much reduced scale. For example, in a test marketing exercise, the experience gained from such activities is useful in determining whether full-scale operation should be implemented. The attraction of such an approach is that the consequences of failure, such as going bankrupt, are avoided but they are relatively expensive forms of experimentation.

In a gaming situation all the features of the system are modelled except for people. A common application of this approach is found in business studies where students are asked to manage a model of a system rather than reality. This approach reduces costs and time while allowing students to get a feel for relationships experienced in practice including the reactions of other people where competitive games are set up. Simulators are also used to convey practical skills as in the operation of large machines or in various skills connected with animal husbandry. The attraction of the approach is that the costs (including the cost of mistakes) and time required to gain experience on real machines and animals is high.

In simulation models instead of people being part of the system, decision rules are explicitly written into the model. This speeds up experimentation so that many more alternatives can be investigated and the implications of time and risk can be explored. The way in which a simulation model traces out a system's behaviour over time is depicted in Fig. 2.7. The input will include a description of the initial state of the system and rules for the number of time periods to be considered. The model will express the relationships between the characteristics of the components as they occur through time. Time may be advanced in two ways: either by updating in uniform intervals of time, months, days or years, or on the basis of the time at which events occur, irrespective of clock or calendar time.

The influence of stochastic variables in simulation models can be captured by random sampling from given probability density functions (see page 79). If this process is repeated many times then it is possible to combine different distributions into one. There are two major problems in this approach. Firstly, the probability density functions are conditional on the level of other variables which may or may not be included in the model. For example, estimates of differences in performance can be derived

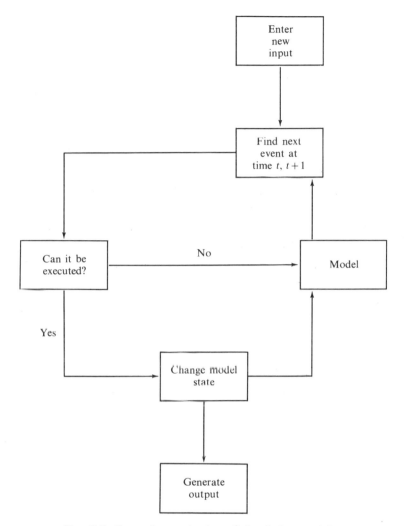

FIG. 2.7. General organisation of simulation models.

fairly easily for different individuals or institutions but this is not the same thing as different performance levels by the same individuals or institutions over time. The second problem of including randomness in simulation models is that the experimenter has no clear idea of the reasons for a particular result. To get round this problem it is preferable to undertake a sensitivity analysis where differences in the levels of variables are investigated systematically.

In analytical models the one to one correspondence between time in the real world and time in the model is usually abandoned. Where this is not the case the size of the model tends to become extremely large. The advantage of analytical models is that they are usually sufficiently simple for a solution to be found using calculus or matrix algebra. Analytical models are usually cheap and can be applied to a wide range of different problems. More detailed models are more expensive both in terms of their construction, their data and knowledge requirement and the computer facilities required to run them. Nevertheless they are at a disadvantage if time, risk, discontinuities, integers and awkward mathematical relationships need to be taken into account. Further, by excluding time they do not provide information on the transitional processes involved by which a system reaches a more preferred state.

Analytical methods provide a mathematical optimum given the constraints and the objectives written into the model. The modeller thus has very few doubts about the answer he gets whereas with other types of modelling there is always a danger that the best answer is not discovered. Simulation models are very useful for testing out 'what if . . . ?' kinds of questions but the search for a best solution is dependent on the model builder asking the right questions. However, this is not to say that the optima derived from analytical methods are any more reliable since they in turn depend on the model builder specifying limits and goals realistically.

PROCEDURES IN MODEL BUILDING[4]

The first step in any model building exercise is to correctly specify the problem. A common mistake is for models to be applied to situations because the model exists rather than because a real problem has emerged. Operational research has classified problems into various kinds including queues, inventories and allocation so that it is possible to use general models provided the problems have the appropriate characteristics. This means that the system under study must be thoroughly researched so that in the formulation of an initial model, the possibility of using the logic and techniques of well-established procedures can be investigated.

Flow diagrams can be useful in the initial formulation of a model. A helpful approach is to start with a highly generalised model which is simple, rather than to try to include all the detail at the first attempt. Refinements can come later once the model builder is sufficiently familiar with the real system to be confident about what elements are important.

A common criticism of modelling exercises is that insufficient data exist to justify the effort. Firmly committed model builders retort that errors in information are a poor reason for making further errors in analysis. Indeed to leave out a variable from a model because data are not available implicitly assumes that the variable has no influence which is probably the least likely situation in practice.[5] Two further arguments are that if a model is primarily designed to demonstrate system behaviour then the actual level of the output variables is not so important. Furthermore, if assumptions have to be made, it is possible to test the sensitivity of the output to these assumptions before steps are taken to acquire more data. If in this way the costs of acquiring irrelevant data are avoided a modelling exercise is useful. It is unusual to find data in the precise form required by a model and normally relationships may have to be transformed or the model reconstructed in order to get round such problems.

The construction of a specific model forces the model builder to systematically describe the system either as a set of equations or as a computer program. This can be a time-consuming task especially for the more descriptive types of model and care must be taken to check that the model gives consistent answers. Negative values for variables may not make sense yet they can easily occur unless steps are taken to prevent this.

Validation

A difficult philosophical problem arises when the 'truth' of a model is tested, especially if a corresponding real system does not exist. Models can be tested by examining the assumptions on which they are based to see if they correspond to everyday experience and well-tested facts and laws. It is often the case that highly sophisticated models deter critics from stepping back and examining the model builder's perception of the real world. Confidence in the assumptions will be increased if these have been empirically tested. In the case of relationships included in models which have been derived in controlled experimental conditions it is difficult to be sure that the coefficients will remain fixed once the controls are relaxed. Thus, an overall check on the behaviour of the model's output compared with historic data from the real system is also required. If time permits, the model can be tested as a predictor of future events in reality. This approach has the advantage over a historical comparison where past data are used both to estimate relationships included in the model and to test the output.

Tests for goodness of fit between the paths of the main variables as predicted by the model and historical data are also a subjective process. Objective statistical tests such as multiple correlation and a rigorous analysis of residuals or unexplained variation only assist the model builder with his judgement. How good a fit is satisfactory, for example? How sensitive is the model output to errors in the coefficients used in the model? Does the model behave like reality in the sense that the predicted timing of direction changes, and the amplitude of fluctuations matches those in the actual system? Simulated behaviour is often easier to model in relative terms than in absolute terms. Good predictive model behaviour is, however, always subject to the possibility of major structural changes in real systems so that the truth of a model can only be transient.

The Use of Models

Experiments with the model can produce a great deal of output and devising means of handling this is an important part of the model building process. It is often the case that experimental designs are used but it should be remembered that the variation in a model is all under the control of the model builder. Errors therefore are more likely to be due to the specification of the model rather than randomness. The problem can be overcome by restricting experiments rigorously to the defined purposes of the model building exercise and by keeping the model as simple as possible.

There are two main ways in which models are helpful to decision makers. They help in understanding systems by providing a means of thinking and drawing the manager's attention to the critical elements which affect performance. The actual output of models is also useful where the accuracy can be relied upon. For example, compounders or farmers who use linear programming to derive least cost rations receive specific recommendations as to which ingredients to use. They also gain an insight into which ingredients affect the price of the ration most and which nutrients are commonly in short supply.

Attention must also be given to the capabilities and information that is readily available to the potential users of models. For example, modern computing facilities enable interactive programs to be written which means that the input to be provided can be acquired as answers to a set of simple questions The design of the output is also important. Is the information really necessary? Can it be easily read and understood? Explanatory manuals help the user to run models and they can form the basis for training and persuasion exercises.

A common experience is that the time and effort required to enable managers to use models is just as great, if not more so, than the energy required to build the model in the first place. Advocates of teamwork and consultation with target users before modelling begins do not recognise that such an approach will not solve the problems of user acceptance by itself. Potential users often do not recognise the power of the modelling process. They find it difficult to specify what their needs are or expect too much too quickly of the model and the computer. Teamwork is fine in principle but to perceive, build, apply, sell and service a model demands a great deal of energy and dedication.[6]

REFERENCES

1. RAUSSER, G. C. and HOCHMAN, E. (1979). Dynamic agricultural systems: economic prediction and control. In: *Dynamic economics: theory and applications*, Vol. 3. Elsevier North Holland, New York.
2. NAUGHTON, J. (1976). *Scientific method and systems modelling. A third level course.* Systems Modelling Unit 8. The Open University Press, Milton Keynes.
3. DALTON, G. E. (Ed.) (1975). *Study of agricultural systems.* Applied Science Publishers, Ltd., London.
4. WRIGHT, A. (1971). Farming systems, models and simulation. In: *Systems analysis in agricultural management.* (Dent, J. B. and Anderson, J. R. (Eds)) John Wiley and Sons, New York.
5. FORRESTER, J. W. (1961). *Industrial dynamics*, MIT Press, Cambridge, Mass., 464 pp.
6. DALTON, G. E. and PACK, B. S. (1981). *The development of a computer model and its application in practice by an extension service,* In: *Computer applications in food production and agricultural engineering,* Proceedings of the IFIP TC Working Conference, Havana, Cuba. (Kalman, R. E. and Martinez, J. (Eds)), Elsevier North-Holland Publishing Co., Amsterdam. pp. 235–43.

CHAPTER 3

The Environment

An environment is generally defined as those influences on a system's performance that can safely be assumed to be outside the operator's control. It is also assumed that the output of the system has no effect on the environment. Three classes of environmental influences can be identified; namely, the physical environment, the economic environment and the social environment. Each of these classes are systems in themselves and are characterised by interdependence and interaction.

An environment will be defined in a different way according to the role of the manager. Referring to Fig. 2.1 a cow may be part of a system or it can form an environment depending on the level of investigation.

THE WEATHER

The weather causes many problems in agriculture. Light, heat, precipitation, wind all combine to determine the regions of the world where particular types of agriculture are feasible.[1] Management is primarily concerned with designing systems which can cope with the weather and in particular with variation in weather events.

Measurements of weather variables are available in most localities. They include rainfall, air temperature and relative humidity, wind speeds and occasionally sunshine hours and measurements of radiation. The data are usually collected on a daily basis and are published in the form of means. The data are more reliable for decision-making purposes if they have been collected for a long period of time. Data collected for 20 years or so are not nearly enough to discern trends and cycles. Van Kampen[2] in a study of harvesting costs on a Dutch polder produced markedly different results when weather data for the period 1931–1949 as opposed to 1949–1967 were used.

The effect of weather depends on the coincidence and timing of events. Rain with high winds and cool air temperatures produces greater heat loss than in warm, still conditions. The pattern of rainfall over time is important for the completion of operations such as hay making and cereal harvesting where a sequence of dry days is required. The cost of extreme conditions such as droughts, floods, gales, heavy snow and frost depends on when they occur. The effects will also depend on the characteristics of the rest of the physical environment such as the type of soil and vegetation.

Management is continually making decisions in anticipation of future weather events. On a day to day basis the following types of problem occur: Will the weather be fine long enough to dry the hay? Are soil conditions suitable for planting? There is an almost continuous balancing of the costs and benefits of action now as opposed to later weighted by the likelihood of favourable and unfavourable events occurring.

A preoccupation of market analysts is to predict the effect of weather on crop yields and total supplies. The coffee market for example is influenced by the incidence of frosts in Brazil, a major producer; the grain market by snow cover, and soil moisture at critical points in the growing season.

Management has to cope with the weather. One way of doing this is to invest in facilities which enable operations to be carried out more quickly or which create an artificial environment. Soils can be irrigated, drained and protected from erosion. They can be left fallow to build up soil water reserves. The main benefit of such practices is that the type of agriculture can be changed. They also reduce the variability of returns, which is especially important where agriculture is carried out in extreme conditions. Records show for example that low average values for rainfall are negatively associated with measures of variance. In other words, the variability of rainfall increases with aridity.[3]

Future weather can never be known for certain. The weather that does occur can be critical to the success of an enterprise. A financially vulnerable business may survive if immediate weather events have a favourable impact on profits. A succession of poor years can precipitate ruin. The same logic applies to the pay-off from projects which are designed to counteract extreme weather events, e.g. floods or drought. Since the returns early in the life of an investment are most important in determining profitability and restoring liquidity, the sequence of future weather events can greatly affect the benefits from such measures.

An important feature of the physical environment is that for the most part deficiencies in information can be overcome by basing estimates on physical laws.[4,5] These estimation procedures can be fairly sophisticated as, for example, in calculating run-off after a rainfall or the rate at which crops dry out. Coefficients may be unreliable but at least the way in which physical systems such as the weather behave is predictable. A similar approach can also be adopted to predict the behaviour of the build-up of pests and diseases. For example, given assumptions about breeding rates, feed supplies and the efficiency of predators and chemical means of control, estimates of the damage done by insects can be fairly precise.[6]

THE ECONOMIC ENVIRONMENT

The understanding of economic systems is far less developed than that for physical systems. Nevertheless some relationships have been observed and validated. Economics in a general sense is about the behaviour of large groups of people; the specific activities of individuals are ignored. Two sets of knowledge are relevant: firstly, the behaviour of agricultural markets, and secondly, the behaviour of economies in general.

Agricultural Markets

Prices in free market economies are the means by which the self-interest of both producers and consumers is reconciled. An upward shift in price means that consumers in aggregate will buy less of a commodity and producers will be encouraged to produce more. In the reverse situation as prices fall consumers seeking to obtain as much satisfaction as possible from their expenditure will buy more and producers will be encouraged to produce less. The pattern of prices over time traces out the conflict between buyers and sellers in the market place. There are longer-term influences on consumption and production such as changes in incomes, population and technological developments.

Purchases of food by people are not simply a matter of satisfying nutritional requirements. Obviously these have to be met if life is to be sustained, but most people most of the time eat for enjoyment. The enjoyment of food reflects prejudices, folklore, associations, custom, tastes and the influence of advertising. As more food is eaten the satisfaction from each extra amount of food declines if only because of the limited capacity of the human stomach. This means that if food prices fall there is likely to be a less than proportionate rise in consumption. For staples in the diet—bread, potatoes, rice, cassava—consumption levels tend to be constant irrespective of price changes. For preferred foods such as meat, fruit, vegetables and for services added to food, the amount purchased is more responsive to price movements. This means that, for a given change in supply, prices will move more for staple commodities than for luxury foods.

Another feature of the food market is that if the prices of goods other than food become relatively more expensive, then for consumers on fixed incomes there will be less funds available to buy food. The usual response to lower disposable incomes is for the mix of food expenditure to change in favour of staples and away from preferred items such as meat and fruit. As incomes rise it has been found that the proportion of expenditure on food declines, reflecting the relatively constant demand for food. However, with rising incomes there is enhanced demand for services attached to food such as packaging, preparation and processing into different forms.

Agricultural producers in the short-term are unable to respond quickly to price changes because of the time it takes to produce or cease to produce extra output. Existing prices are therefore unreliable indicators of what to produce or how much to produce especially for products with a long lead time such as tree crops and beef. Once a crop is established or a breeding herd built up then production is almost inevitable. Prices would have to fall to extremely low levels before it would not pay to harvest a crop or slaughter an animal. Thus while existing prices are perhaps the best available evidence to producers of what their future production plans should be, experience shows that prices do not remain constant. It is almost true to say that the least likely price in a year's time is today's price.

Expectations of future prices need to take account of other information available to producers such as stock levels, crop reports, planting intentions, chick placings and weather events, together with an understanding of the way the market works. Confidence is required to invest in production and it can easily be destroyed. Producers do over-react to shortages and surpluses because confidence is too high or too low. It also reflects the fact that there is normally a shift in bargaining power between buyers and sellers according to the level of supply. When shortages occur producers are in a strong position. When surpluses abound the situation is reversed.

Long-term supply levels will only be maintained if costs, as measured in terms of the use of resources in other ways, are recovered and there is sufficient reward to compensate producers for the risks involved in production. The risks are substantial in agricultural markets since the short-term inflexibility of supply and demand can produce marked and sudden price movements. Price falls will result in high cost producers leaving the industry. High costs need not necessarily be associated with adverse physical environments. Mistakes in management, bad luck, poor investment decisions and wrong choices can be equally costly. As stated previously, costs measured in an accounting sense are misleading. Production may continue in a poor physical environment because the land, labour and capital cannot produce a greater return in some other activity. It is therefore dangerous to base agricultural policy on biological measures of efficiency.

One result of an unregulated market in food is that if prices are high then the costs of reduced consumption may well fall upon the poorest section of the community. In many countries this is morally unacceptable so that compensating mechanisms are employed such as price subsidies, rationing and free food for recognisable groups such as children and the old. This problem is also an international one. The best way for an individual nation to ensure security of food supplies is to be rich, in so far as it will be able to afford to buy food even when prices are high. High prices will periodically occur especially for staple commodities and irrespective of general trends.

High prices are even more likely because of the residual nature of the world markets for most commodities. International trading agreements and Government measures such as import levies, domestic price supports and quotas ensure that the impact of unexpected events falls entirely on a small proportion of total trade. The problems of adjusting to both surpluses and shortages are therefore exaggerated. The result is that the prices at which trade takes place do not accurately reflect the balance of supply and demand and it will fluctuate considerably. The costs of fluctuating prices will fall most heavily on the poor both directly when prices are high and indirectly because of the inaccurate signals given to both domestic and exporting producers.

A completely free market also ignores the wider costs of changes in the level of production. Marginal producers forced out of business may not find profitable alternative ways of using their resources. They may also suffer substantial capital losses because high incomes in agriculture tend to bolster capital values of land and breeding livestock and increase the rent paid for resources. The reverse happens when income levels fall. Thus in many

countries transfer payments are made to the less privileged producers especially as the main beneficiaries of enhanced efficiency
are consumers. Transfer payments can be designed to assist in the
adjustment process by improving access to new opportunities
through training and re-location and in the provision of
alternative forms of employment. However, the most common
means of supporting the incomes of farmers in rich countries is by
various forms of price support.

Price Support
The problem of agricultural markets is that a slight degree of
under- or over-supply can have a dramatic effect on prices. Prices
in turn affect the welfare of both producers and consumers. For
producers the difficulty of transferring resources to other parts of
the economy can produce a chronic situation of over-supply. The
effect of high prices on poor consumers is particularly severe. The
political response in rich countries is to raise prices that farmers
receive irrespective of the market situation and often at a high cost
to the total economy, to consumers and Government finance. The
debate about price becomes a political matter and statements
on agricultural policy speak of food security, guaranteed supplies, reasonable returns to producers and fair prices to consumers. These objectives conflict but most governments find
that price support is the most acceptable way of reconciling
them.

Price support has some serious side effects. It is a blunt
instrument for meeting social needs. If market prices are artificially raised the poorest consumers suffer most because they
spend a greater proportion of their income on food. If prices are
maintained at a level which encourages overproduction, more
resources are used in agriculture. These resources could often be
better employed in other parts of the economy. Surpluses can
disrupt international markets if they are exported at subsidised
rates. Marginal producers are not encouraged to adjust to the true
state of the market while successful producers can be supported by
policies designed for their less fortunate colleagues. This results in

these farmers making very high profits and acquiring large amounts of land and other forms of capital.

The Wider Economic Environment

Agriculture is part, and in many countries a substantial part, of the total economy. The prosperity of agriculture therefore depends on the health of the whole economy.

Unfortunately, the level of general economic activity does not necessarily coincide with the full employment of all the resources within an economy.[7] Thus, many economies suffer unemployment and inflation. Inflation occurs when demand exceeds supply and consumers require of the economy more than it is capable of producing. A recent worrying feature of economic performance is that inflation and unemployment can occur simultaneously. Some structural problems exist when this situation arises but its causes and cure are unclear.

Assume a self-contained economy, that is one that does not trade with other economies. Within such an economy the needs of consumers will be met by the goods and services produced by domestic firms. The revenue of the firms is in turn generated by the expenditure of consumers so that we can think of two ways of measuring the output of the economy, namely, the value of the goods and services produced or by the level of total expenditure. The incomes of consumers will be derived by payments for their labour and other resources which in turn will be paid for by the firms from their sales to these same consumers. The economy can be thought of as a closed system. Receipts by firms enable them to pay for the resources provided by consumers which in turn enable the consumers to buy the goods produced by the firms and so on. However, if there are leakages in the system then eventually the level of economic activity would run down and approach zero. Leakages can occur in several ways, the most common being through savings made by consumers or, if trade is allowed, by importing goods and services from other economies. Economic activity can be raised by injections of payments into the system through investment by firms and by them selling exports to customers in other economies. The problem is that the leakages from the system, i.e. savings and imports, need not equate with injections into the system, namely investments and exports, so that at any given time an economy can either be expanding or contracting. Thus to maintain unemployment at low and stable levels and to avoid severe inflation, the economy, like any other system, has to be managed and the way this is done is through government intervention.

Two broad policies are open to government. The first one is monetary policy which is based on the identity that the price of goods multiplied by the amount of goods in an economy should equal the amount of money in circulation times the velocity of money in circulation. The belief is that by controlling the amount of money then automatically the price and quantity of goods will adjust. Unfortunately it is very difficult to control the amount of money in an economy. One way is to vary the rate of interest but this has serious drawbacks including the possibility of destroying business confidence. Investment and saving plans will only change significantly for large changes in interest rates. The amount of money can be regulated by direct controls on the banking system but it has been found that as soon as one part of the system is controlled another part takes over its role.

The second means by which governments control the general activity level within an economy is through taxation and government expenditure. Government expenditure will increase the demand for goods and services. It can be selective by, for example, the issuing of grants for particular types of investment. Taxation is a leakage in the system since it reduces effective demand. Unfortunately, the political acceptability of fiscal policy runs into the erroneous yet popular belief that governments, like individual households, should balance their budgets by equating expenditure with the revenue from taxation. The whole purpose of fiscal policy is to counteract the tendency of the economic system to find a balance without a sensible level of resource use. This will require, in times of low economic activity, budgetary deficits and in times of excess demand, budgetary surpluses.

Ideally, fiscal policy is counter cyclical. In modern economies the flexibility of fiscal policy can be reduced by committed

Government expenditure on social programmes. Increases in taxation are not particularly popular at the time of elections so that short-term political expediency can result in the mismanagement of the economy.

The International Economy

The same principles of economic management also apply to the maintenance of international economic activity. The expenditure of nations is matched by their receipts. One nation's balance of payments surplus is another nation's deficit. An individual economy which restricts imports artificially as a means of reducing leakages within its own system, reduces the stimulus to activity in the economic system of the country supplying those imports. 'Beggar my neighbour' policies can thus result in a serious curtailment in the overall international level of economic activity. This of course has been realised and there are international agreements which try to prevent this from happening by, for example, reducing barriers to trade. International bodies such as the International Monetary Fund have also been set up which enable individual countries to finance their deficits by borrowing.

In practice, international control is often achieved by the actions of dominant economies such as that of the United States of America. If the large American economy runs a deficit on its trading account with the rest of the world, this provides a powerful stimulus to the economies of its trading partners. Likewise if the American economy grows this too is likely to have a significant effect on the economies of its trading partners. On the capital markets the large savings by OPEC countries could easily have resulted in massive world deflation had the international banking system not recirculated these funds fairly successfully. The OPEC countries have also initiated large investment programmes which have moderated the potentially large leakage effects of their large earnings.

The Interaction between Agriculture and the Rest of the Economy

The performance of the total economy is just as important for agriculture as it is for other industries. Prosperity, or lack of it, has a direct income effect on the amount and pattern of food consumption and for commodities such as wool there is a very close relationship between consumption and general economic activity. A depressed economy means that adjustment is more difficult since there is a lack of alternative employment opportunities for surplus labour within agriculture. Governments may not be able to operate any form of price support or to develop infrastructure such as roads, irrigation, research, credit, health and education. Crucial supplies may not be available because of foreign exchange and production problems. It is no accident that support for agriculture is mainly a pre-occupation of rich countries.[8]

One of the features of unsuccessful economies is inflation and this can have a devastating effect on capital intensive enterprises such as agriculture. Consider the simple case of a machinery dealer selling machines. Suppose the machine is bought for £100 and is sold some time later for £200. If expenses amount to £50 then a profit of £50 is left. Assume half this profit is paid out as taxes and dividend leaving £25 for reinvestment. Now, if, because of inflation, the next machine to be purchased costs £130, the business is faced with a financial problem since there is insufficient cash being generated by the business to pay for the replacement. The only way out of this problem is either to borrow more funds or to contract the size of the business.

Other features of unsuccessful economies consist of unemployment, large balance of trade deficits and rapid exchange rate depreciation. All are interdependent and can influence agriculture to a large extent. Unemployment in urban industries can result in government policies which keep food prices unrealistically low. Taxes on agricultural exports may be imposed to finance the import of food products which are then subsidised without regard to the long-term detrimental effects on the local agricultural industry. Direct alleviation of unemployment may result in 'back to the land' policies through resettlement schemes which are often unrealistically expensive per job created. Balance of trade figures are often used as an argument to support domestic agriculture. Falling exchange rates, by making imports more expensive and exports cheaper, tend to favour domestic agricultural production. Unfortunately, for many developing countries who rely to a great extent on the export of one or two commodities to earn most of their foreign exchange, falling exchange rates mean that a greater volume of exports is required to purchase a given quantity of imports. If the market for their products is limited then expansion of output to compensate for this will drive down prices more than proportionately.

Autarkic policies are often popular solutions to both general economic problems and agricultural problems. Self-sufficiency would require that the resources required by agriculture are also domestically produced and this may not be feasible because supplies may not exist or it would be prohibitively expensive to produce them locally. The policy is also blinkered in that it denies countries, both rich and poor, the opportunities to exploit the large increases in technical and economic efficiency that can be tapped through trade. It makes both biological as well as economic sense to grow sugar in tropical climes and to raise ruminants for milk and meat production on natural grasslands which cannot easily produce cereals. The only way in which many developing countries can earn foreign exchange is to export agricultural products.

Price support systems which result in dumping, export subsidies and quotas create disorder in world markets and distort trading patterns in agricultural products. Inappropriate 'food aid' to poor nations can be disruptive to local agriculture. In the short term, prices are lowered which acts as a disincentive to local production. In the longer term, tastes are developed for commodities which cannot be produced locally, e.g. wheat and dairy products. Food aid can of course be helpful but it needs to be planned according to the objectives of the recipients rather than as a convenient way of disposing of accidentally produced surpluses.

The importance of the general economic environment and the way it affects and is affected by agriculture is illustrated at the time of writing by the large imports of grain by the USSR. These imports are being partly financed by the sale of gold and diamonds with a dramatic downward effect on the world price of these commodities. Extended credit terms are being sought by the USSR along with Eastern Bloc countries. The scale of these imports has contributed to the world price of grain being higher than it would otherwise have been given that the world's major exporter, the USA, has had a record harvest. This has reduced the size of the subsidies paid on the surpluses exported out of the European Economic Community which in turn has reduced the internal pressure for budgetary reform of the Common Agricultural Policy (C.A.P.). However, the external pressure to reform the C.A.P. is building up because the Americans are losing traditional markets for their agricultural products to subsidised European supplies. A trade war is threatened between the USA and the EEC which could include other commodities in oversupply such as steel and chemicals.

The economic policy of the American Government is also causing concern. A projected budget deficit coupled with a tight monetary policy, it is believed, will result in continued high interest rates. The effect will be to raise international interest rates as capital is attracted to the United States. This will also sustain the exchange rate of the dollar against other currencies. Since oil and grain are normally paid for in dollars, the inhabitants, including farmers, of the countries of the world who buy substantial quantities of these commodities will face higher bills than they otherwise would as well as higher interest payments.

SOCIAL ENVIRONMENT

The social environment for agriculture is a good deal more localised than international finance. Nevertheless it is growing in importance. There are serious worries about the costs to society as a whole rather than the strictly commercial costs of modern agricultural practices. The best example of this is pollution from chemicals used in agriculture including pesticides in crops and growth promoters in animals. In richer countries there are worries about the countryside as habitats for flora and fauna are being removed and there is general concern about the breakdown of natural control mechanisms as a result of carelessness, ignorance and commercial pressures.

One of the more serious concerns is the effects of change in agriculture on those people who work within it. What is the point of producing landless labourers alongside rich, powerful, landowning but highly productive large farmers? What is the effect on the minds and spiritual welfare of people working in intensive livestock units? Indeed what are the effects on the welfare of the animals? It is sometimes forgotten amid waves of sentimentality over animals that cruelty and exploitation of animals degrades human beings.

Modernised agriculture tends to produce rural depopulation with a consequent decline in the services available due to higher unit costs of providing for fewer people in rural areas. The fate of the migrants causes serious problems throughout the world. This is of particular concern in developing countries since people seem to drift to the towns because they perceive a greater chance of enjoying a higher standard of living which includes the enjoyment of public utilities.

There seems to be an impatience or ignorance of some of the most profound value systems that exist within agricultural communities. One of them is the feeling of stewardship over the land and other resources. Another is the value of human activity and experience for its own sake rather than purely for the satisfaction of economic objectives. Of course rural life can be idealised especially by those who can afford it. All societies in the end have to decide on satisfactory means of feeding themselves.

CONCLUSIONS

The relevant environment for agriculture is as wide as the entire range of human activities and experience. It includes the weather, the physical and biological environment, world events, economic and social systems. They can all influence the performance and importance of agricultural systems, often in unexpected ways. For example, how will new developments in biology and electronics change agriculture? How quickly will these changes occur? The ability to comprehend and to put into perspective new situations and to make sense of change is an important and necessary skill for managers of agricultural systems. However, these skills, while necessary, are not sufficient. The purposes of agriculture must be kept in their proper place. Material things are only a means to an end and confusion about ends makes management irrelevant. Agricultural systems can only be operated within the context of the highest environment of all, that of truth and purpose.

REFERENCES

1. DUCKHAM, A. A. and MASEFIELD, G. B. (1970). *Farming systems of the world*, Chatto and Windus, London.
2. VAN KAMPEN, J. H. (1969). *Optimising harvesting operations on a large-scale grain farm*, PhD Thesis, Wageningen.
3. MCINTIRE, J. (1981). *Food security in the Sahel. Variable import levy, grain reserves and foreigh exchange assistance*, International Food Policy Research Institute. Research Report 26. Appendix 1.
4. MONTEITH, J. L. (1973). *Principles of environmental physics*, Edward Arnold, London.
5. VAN ELDEREN, E. (1977). *Heuristic strategy for scheduling farm operations*, Centre for Agricultural Publishing and Documentation, Wageningen.
6. CONWAY, G. R., NORTON, G. A., SMALL, N. J. and KING, A. B. S. (1975). A systems approach to the control of the sugar cane froghopper, In: *Study of agricultural systems*. (Dalton, G. E. (Ed.)) Applied Science Publishers, Ltd., London.
7. DONALDSON, P. (1973). *Economics of the real world*, British Broadcasting Corporation and Penguin Books, London.
8. MOLLETT, J. A. (1982). Agricultural investment and economic development — some relationships, *Outlook on Agriculture*, **11** (1), 27–32.

CHAPTER 4

Planning

A DEFINITION OF PLANNING

Planning is the contemplation of possible activities before a choice is made in the hope of ensuring a more acceptable result. Consider the planning of an ordinary day for the type of person who has some choice about what he does. There will be a long list of things to do but insufficient time or energy available to do everything. Priorities have to be set according to the level of anticipated rewards or penalties from doing or not doing different things. Working out the consequences of different activities is not a simple matter since there is continual conflict between daily and longer run priorities, and decisions which may appear to be wise in the present can be nullified by future events. Consequences will also be modified as a result of the interactions between different decisions made over time. A good or bad decision can be undone or improved by a subsequent decision made in the future.

The planning process can be split up into several parts. A manager can only plan if he defines his objectives. It is only when his objectives are not being met that dissatisfaction will be experienced and change will be contemplated. The choice of what to do will depend on the way in which different plans fulfil his goals.

The quality of decisions about what to do depends too on the success achieved in thinking up different ways of doing things. Imagination is required to think of ways of overcoming restrictions on action such as finance, capacity bottlenecks, skills and competition. There is always a danger that the best solution is not thought of or is overlooked. Once different courses of action are specified it is necessary to predict their likely consequences under different likely sets of future conditions in order to choose between them.

INFORMATION

At all stages in the planning process information is required which takes time and resources to collect and can result in delayed decisions. Decisions often have to be made in haste by a given time so that planning inevitably takes place with less knowledge than is potentially available. Extra information can cost more than it is worth.[1] Information, for example, that does not change a decision has no value. In the following example, in Table 4.1, a farmer can apply three levels of nitrogen to his crop, high, medium and low, the response to which will vary according to the weather, good, fair and poor. If the likelihood expressed as probabilities of the three types of weather are 0·2, 0·4 and 0·4 respectively, the expected or average margin for each fertiliser strategy can be calculated.

TABLE 4.1
Margins According to Fertiliser Level and Weather

| Fertiliser level | Weather type and likelihood | | | Expected margin |
	Good 0·2	Fair 0·4	Poor 0·4	
High	£200	£140	£0	£96
Medium	£150	£120	£40	£94
Low	£100	£80	−£20	£46

The highest average margin is obtained by pursuing a high fertiliser policy. If it were possible to purchase a perfect weather forecast, the average margin would become 0·2(£200)+0·4(£140)+

0·4(£40)=£112. Because the forecast alters the fertiliser strategy when poor weather occurs from high to medium the gain of £16 (£112 − £96) represents the value of the forecast.

The purchase of less than perfect information which changed the estimated probabilities to 0·25, 0·35, 0·4 would produce a different set of expected values, namely 109, 79·5 and 45, in which case the best policy to maximise the expected margin would still be the high fertiliser policy. This information would not change the decision and therefore has no value.

OBJECTIVES

The importance of objectives is always stressed in discussions about management because choosing between different activities depends on what is wanted. Managers and society have conflicting needs which change with circumstances. Failure to satisfy physical needs produces easily defined discomfort such as hunger, tiredness, fear and sickness. Social needs such as a good reputation and a place in society are measured by less dependable feelings such as emotion and a sense of belonging but can be powerful motivators. Likewise spiritual needs such as creativity, purpose and righteousness reflect the value judgements of the individual and the society to which he belongs. Objectives, then, are highly personal, often vague and are often not met.

It is unusual for objectives to be stated explicitly. Rather, a manager experiences or anticipates dissatisfaction with his lot. The seriousness of his situation depends on the extent to which objectives are not being fulfilled and how important these objectives are. For example, a person who is hungry with no money to buy food and who refuses to steal it ranks his sense of what is right or is acceptable in society as more important than his own physical requirements. However, the extent of both his hunger and his fear of either going to jail or eternal damnation also determines his action. He may become, in the extreme, so hungry or indignant that ultimately his sense of righteousness and fear of punishment are insufficient to prevent him from stealing. More usually, provided different courses of action are regarded as being socially and morally acceptable, there will be a trade-off between objectives such as more profit or being better fed and other social activities which lead to, say, more leisure.

It is not possible in most situations for people to state explicitly how much more important one objective is than another or to be able to quantify different degrees of satisfaction of one objective in terms of some other objective. For example, different degrees of hunger cannot be measured in terms of, say, honesty unless a common scale of measurement can be found. In the absence of such a common measure, choice cannot be reduced to an arithmetic exercise. Even if it could, it would take too much time and effort for most people to specify their respective preferences in terms of each other.[2]

A helpful way of explaining how the human mind might balance different objectives is the idea of ranking objectives in order. Some objectives may take the form of absolutes or 'must' objectives. If, for other objectives, at any particular moment a decision maker can say which objectives are more important to him, even though he cannot say by how much, then alternatives can be screened in the following way.

First actions are examined in the light of absolutes. Provided they do not break *must* criteria, then alternatives can be compared in terms of how well they meet the most important non-absolute objective. If two alternatives have the same performance at this level, then they are compared at a lower level and so on until one alternative out-performs the others. This lexicographic approach is the same rule as is applied in ordering words in the dictionary.

Another possible way for people to think is to weight non-absolute or secondary objectives by an ordinal ranking of satisfaction and to choose the option with the highest ranking. Thus, motor cars may be ranked against each other by their aggregate score for such factors as colour, interior decor, luggage capacity, comfort and so on.

Degree of Satisfaction

How does a manager rank his objectives in order of importance?

The simplest situation is where an objective is either met or is not met, the binary case. This type of situation applies best to concepts of right and wrong. Many objectives have the characteristic that they are more or less satisfied, such as being more or less hungry and having more or less status. At any point in time we can imagine the manager not achieving satisfactory levels of attainment for a few of his objectives and thus at that time he will rank these objectives as being more important than others in deciding what to do. This situation is depicted in Fig. 4.1 where a continuum represents the range of satisfaction levels.

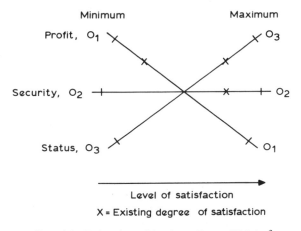

FIG. 4.1. Balancing objectives. From Wright.[3]

The manager has three objectives, Profit O_1, Security O_2 and Status O_3, with minimum and maximum limits. For objectives O_2 and O_3 he is achieving in his current activities high levels of satisfaction but a low level of satisfaction O_1. Thus, in endeavouring to improve his situation the manager will be interested in courses of action which will shift O_1 to the right, provided by so doing no limits are broken and the degree of satisfaction for O_2 and O_3 is not reduced by too much. In the end there would seem to be no way out of measuring the comparable objectives against each other. Is a substantial increase in O_1 worth a slight reduction in O_2 and O_3? All the decision maker has to do is to say that a new combination of levels is more or less preferable to the existing state which is perhaps not quite so difficult.

Objectives differ between people and a satisfactory level of attainment for one person does not mean that it is so for another. This means that decision making is a highly personal matter. This creates problems in explaining how groups of people make decisions collectively. Collective decisions could well rule out the unusual, the imaginative or the risky alternative if a consensus has to be reached.

Definition of Objectives

The will to live is the most basic force governing the characteristics of agricultural systems. Survival for the human race depends on enough food being available and meeting this objective with the resources that are available is the prime function of agriculture. It is a *must* objective. Other absolute objectives are also connected with survival. Individual firms and institutions strive to survive competition and must in the long term at least cover their costs. There are more simple constraints such as that the environment must not be harmed through, for example, soil erosion, disease or pollution, and people and animals must not be subjected to cruelty. Food must be safe to eat and palatable and be continuously near to consumers in suitable amounts at a price they can pay.

Social objectives also influence the conduct of agricultural systems, especially the concept of equity both for consumers and producers. Policy statements on agriculture by governments usually stress fairness in the pursuit of feeding all the population and ensuring adequate rewards for producers at fair prices for consumers. All businessmen work within codes of practice which may or may not be reinforced by the law. Conventions such as the right to cultivate, property rights, terms and conditions of trade, are all necessary for agricultural systems to function smoothly and efficiently. Very large transactions are agreed on trust. Higher values still are reflected in the sanctity accorded to animals, the land and even the practices carried out in agriculture, and the whole industry is marked by individuals possessing a strong sense of trusteeship, service and conservation. People within agriculture are of course influenced by the activities of the rest of society. Practices and attitudes are infectious, changes occurring asymmetrically, as the actions of innovators are copied by their fellows.

Within these constraints there is room for manoeuvre. People can within the limits of their appetite eat more food. They can buy types of food such as meat and fruit which they prefer to staples such as carbohydrates. More or less services can be added to the food including processing and packaging. The reliability of supplies can be improved by increasing the amount of stocks or by improvements in production methods and distribution. Prices can be influenced by subsidies on both production and consumption.

Agriculture, the process by which people are fed, cannot be regarded as simply meeting nutritional needs. People also set constraints on the way agriculture operates according to their beliefs on fairness, their need for security and their feelings about what is pleasant, and moral. An important objective which normally dominates these factors is that their requirements from agriculture are achieved if not at a minimum cost then certainly at an acceptable level of cost.

Sub-objectives

The lexicographic ordering of objectives also includes the situation where general objectives can be broken down into sub-objectives that need to be fulfilled in order to achieve the higher goal. The easiest way for farmers, for example, to improve receipts in a situation of fixed product prices is to produce more physical output by raising yields. Thus, in most situations a suitable criterion for improving profits is to seek higher yields especially as it usually costs very little more to produce a good crop or a prolific animal. A more general example of the way in which results as measured by return on capital can be built up from an array of contributory factors is shown in Fig. 4.2. It is evident that several of the factors contribute in a variety of ways to both the overall criterion and to other sub-objectives. The network does not allow the degree of interdependence to be represented accurately. A better approach might be a circular diagram as

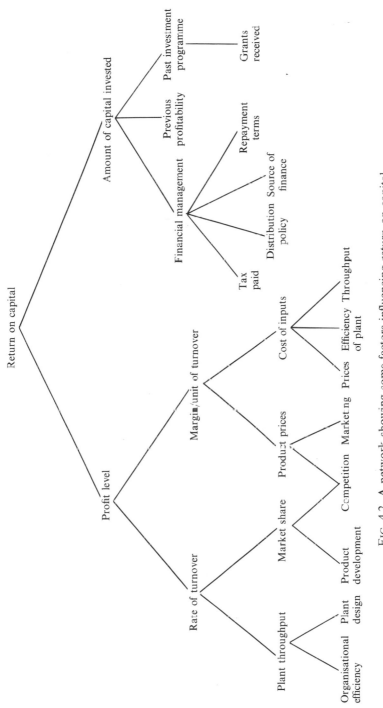

FIG. 4.2. A network showing some factors influencing return on capital.

suggested by Spedding[4] but even this would not allow time dependence to be illustrated.

The analysis of 'what matters' is the key to success in the management of agricultural systems.[5] Identifying what is most important does, however, require an overall view of a system as well as a clear understanding of the interaction between different objectives at the same and different levels. A major problem, for example, in many countries is the distribution of food to poor people. This might be tackled through a system of consumer subsidies but more effective progress might be made by increasing supplies of food through stimulating production or trade and hence reducing prices. Food self-sufficiency is another common goal as a means to improve security. In many circumstances encouraging local production can prove to be extremely costly which will in turn reduce the ability to pay high prices when supplies are short. Resources that are required to produce food may be just as liable to failure as food itself.

Profit as an Objective

Most planning models simplify reality by assuming that profit is the only criterion a decision maker considers and furthermore that he is out to make the most profit possible. This is not how reality is but the assumption will be reasonably valid in situations where profit is complementary to other important objectives such as status, security and survival. Where competitive objectives are preferred to profit a helpful piece of information for decision making is the cost of these objectives in terms of profit forgone. Achieving the highest level of profit may be a residual rule once a given level of other objectives has been made. However, constrained or safety first profit maximisation ignores the possibility of some trade-off between profit and other objectives.

The term 'profit' needs careful definition. Several types of income and expenditure may not be measurable in cash terms. Costs such as depreciation may be incorrectly calculated and errors in changes in stock valuations can seriously distort estimates of income and expenditure. A system's view of what profit is would include costs and returns to the community or business as a whole rather than to individuals or units within them. These 'externalities' would include the cost to society of pollution and structural change such as the plight of displaced workers. Market prices are often inappropriate guides to overall costs so that it is necessary to apportion 'shadow' prices to such resources.[6] 'Shadow' prices for items such as labour, capital and foreign exchange are normally calculated on the basis of their earning capacity in alternative activities. These are usually much lower than market prices for labour and much higher for foreign exchange and capital.

Time Preference

Most people prefer to acquire satisfaction in the present rather than to wait for it. Waiting creates uncertainty and adds to costs in that the freedom to use resources in other ways is not available. Profits can only be spent, borrowed against, or invested if they have been received. One way of allowing for this is to discount future costs and returns.[7] The discount rate may allow for interest, risk, inflation, taxation and the time preference for money. In this way alternative opportunities can be compared in terms of a common Net Present Value. A variant of the approach is to work out the breakeven interest rate. This is sometimes called the Yield or Internal Rate of Return and it measures that discount or interest rate which a stream of net benefits over time would just cover after repaying the initial capital. A simple illustration of the principle is as follows. Suppose £100 is invested and is expected to produce in its three year life an additional cash surplus net of extra cash costs of £40 per year for two years and £40·70 in the third year. The pattern of cash balances is as follows:

TABLE 4.2
Calculation of the Interest on Outstanding Capital in an Investment Project[a]

Year	Owed at beginning of year	Interest charge at 10%	Net cash flow	Owed at end of year
1	100	10	40	70
2	70	7	40	37
3	37	3·7	40·7	0

[a]Net Present Value at 10% = 0; yield = 10%

This contrived example shows that at the end of the project's life all the initial capital is repaid plus interest on the outstanding capital. Thus the Net Present Value at 10% interest rate would be zero which would also mean that the Internal Rate of Return is 10%. Lowering the interest rate would leave a positive amount at the end of the project life (terminal value) and this could be expressed as a positive Net Present Value.

The comparison of projects purely on the basis of Net Present Value or Internal Rate of Return can be misleading where projects have different anticipated lives. In the following example, while project A has a higher Net Present Value than B, it takes six years to earn it whereas project B has a life of four years. The Net Present Value earned per year of project life is greater for B than A (see Table 4.3).

TABLE 4.3
Evaluating Projects with Different Lives

| | Project | |
	A	B
Net Present Value at 10%	£1 400	£1 300
Expected project life (years)	6	4
Equivalent Annual Value[a]	£ 322	£ 410

[a]i.e. a constant sum of money received for each year of the project life that produces the same Net Present Value.

A similar type of problem occurs when projects involve different amounts of investment expenditure. In this case the return generated on the *extra* capital in the more expensive project is a more appropriate guide for choice than simply looking at the rate of return on the total amount of capital in each project. In the example shown in Table 4.4 the return on the incremental capital in Project B is only 10% compared with 21% in Project A. This incremental return might be lower than could be obtained from investment in other directions.

The time preference for money is an important concept for understanding both the amount and type of investments that are

TABLE 4.4
A Comparison Between Two Projects Requiring Different Amounts of Initial
Capital

Project	Capital cost	Annual net cash flow	Project life	Net present value at 10%	Internal rate of return (%)
A	500	100	10	114	21
B	800	150	10	121	13
B − A	300	50	10	7	10

undertaken by an agricultural community. If high interest rates are chosen for reasons of risk or preference for consumption then only the most profitable projects will be chosen and less investments will be made. Further, really long-term projects will be neglected since the present value of money in the distant future rapidly approaches zero at even modest interest rates. The size of the returns in the first few years of an investment's life are most important in determining actual profitability.

Uncertainty

Uncertainty about the consequences of decisions means that managers have to reconcile their hopes of gain against their fear of loss. There are three elements in this balancing act. Firstly, the extent of the loss or gain; secondly, the decision maker's assessment of the likelihood of occurrence of important events; and thirdly, the satisfaction derived, both positive and negative, from a given level of gain or loss.[8]

If decisions were made according to size of anticipated outcomes weighted by their likelihood then managers would be unable to state which of the situations presented in Fig. 4.3 they preferred.

Conservative managers would rule out actions which would have even a slight chance of ruining them. Other managers or the same managers in different circumstances may relish a gamble. The response to the cases depicted in Fig. 4.3 in so far as they indicate how people react to real situations will depend on the type of individual and his situation. A person choosing £500 for certain in Case I or the bet with low stakes in Case II probably

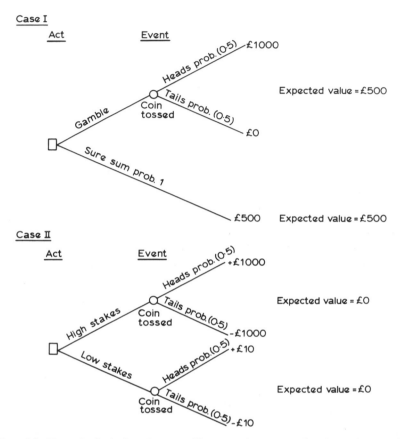

FIG. 4.3. Hypothetical situations to illustrate the expected value criterion for choice. Case I, a choice between a sure sum of £500 or a gamble depending on the toss of a fair coin with prizes of £1000 or £0. Case II, a choice between two gambles depending on the toss of a fair coin with prizes of +£10 and −£10 and +£1000 and −£1000, respectively.

values money in a non-linear way. The first £500 is worth more than a second £500. The prospect of gaining £1000 is not balanced out exactly by the prospect of losing £1000. A graph of the satisfaction or utility derived from different levels of money losses and gains is shown in Fig. 4.4 for a hypothetical individual. Other configurations are possible. For this particular case money gains first of all produce additional satisfaction at an increasing rate but this is not sus-

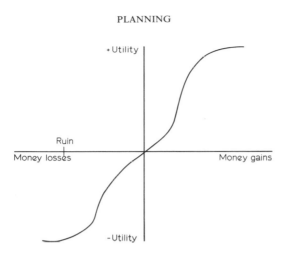

FIG. 4.4. The relationship between money and utility.

tained at higher levels when extra amounts of money add progressively less to utility. Initially for money losses the slope shows that negative satisfaction grows more quickly than proportionate amounts of money but beyond a point of ruin the level of negative satisfaction does not increase.

SEEKING ALTERNATIVES

The main feature of agricultural development is the way in which limitations at all stages in the industry are overcome. Wet soils are drained, dry ones irrigated, nutrient deficiencies are corrected, perishable food is stored, labour shortages are overcome by mechanisation, new markets are found and so on. All these developments are thought of by someone. Every agricultural manager will be restricted in what he can do. He will have limited resources and market opportunities as well as limited knowledge and experience. These restrictions are not usually inflexible, especially in the short run, so that defining the extent of restrictions and ways of stretching them out is a necessary part of management.

A common problem is that of insufficient capacity especially at certain times of the year. The capacity of processing plants,

TABLE 4.5
The Importance of Repayment Terms on Cash Flow for a Loan of £2000 Borrowed at 10% Interest

Case 1—10 year repayment terms

The repaid capital and interest payments on outstanding capital for the 10 years become:

	Year									
	1	2	3	4	5	6	7	8	9	10
Loan repayments	200	200	200	200	200	200	200	200	200	200
Interest	200	180	160	140	120	100	80	60	40	20
Total	400	380	360	340	320	300	280	260	240	220

Suppose the anticipated 'extra' profit generated by the investment is thought to be:

Year									
1	2	3	4	5	6	7	8	9	10
−1500	100	500	500	500	500	500	400	200	100

The net additional cash flow after capital and interest payments is then:

	Year									
	1	2	3	4	5	6	7	8	9	10
Loan	+2000									
Extra profit	−1500	100	500	500	500	500	500	400	200	100
Capital repayments +interest	400	380	360	340	320	300	280	260	240	220
Cash flow	+100	−280	+140	+160	+180	+200	+220	+140	−40	−120
Cumulative cash flow	100	−180	−40	+120	+300	+500	+720	+860	+820	+700

Case II—3 year repayment terms

The cash required for servicing the loan over 3 years becomes:

	Year		
	1	2	3
Loan repayments	666	666	666
Interest	200	133	66
Total	866	799	732

The cash flow position assuming the same level of extra profit is:

	Year									
	1	2	3	4	5	6	7	8	9	10
Loan	+2 000									
Extra profit	−1 500	100	500	500	500	500	500	400	200	100
Capital repayments +interest	866	799	732							
Cash flow	−366	−699	−232	500	500	500	500	400	200	100
Cumulative cash flow	−366	−1 065	−1 297	−797	−297	+203	+703	+1 103	+1 303	+1 403

machines, labour and finance can be the most critical determinants of what and how much is produced. However, such limits can be extended by a certain amount of organisation. If processing facilities are fixed then out-of-season production can be stimulated, plant can be worked longer and more efficiently by reducing breakdowns and waiting time. Identifying the determinants of capacity and treating the whole system as one can be very helpful. Two examples one for finance the other for physical capacity illustrate this point.

Finance

Finance is invariably a restriction on the development of individual businesses. A superficial view might be that in order to return the highest profit capital should be acquired at the lowest possible rate of interest. However, in terms of the availability of cash, repayment terms are also important. Consider an investment financed by a loan of £2000 at 10% interest rate on outstanding capital but in Case I with repayment over 10 years in equal instalments and in Case II with repayment over three years in equal instalments (see Table 4.5).

Comparing Case I and Case II shows that the cumulative extra end of project cash flow is much greater for the short term loan, but the variability of the cash flows is much greater with credit resources being strained most severely in the third year. Further it takes longer to get out of the red, a positive cumulative cash balance occurring first of all in Case II in year six and in Case I in year four.

Suppose a rule was instituted which said that further funds would not be borrowed until a positive cumulative cash flow was achieved. It is apparent that in Case I, growth would be much quicker since there would be a chance to make a new investment in year four, two years before it could take place in Case II. This could well more than compensate for the greater profitability of the shorter-term loan.

Matching Capacity

Capacity in sequential operations needs to be matched as closely as possible if costs are to be minimised. In Fig. 4.5 the area of each square represents the capacity of say separate items of machinery operating in sequence in a factory.

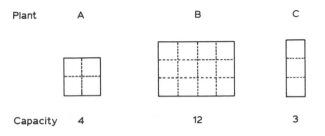

FIG. 4.5. Matching plant capacity.

The ideal arrangement to utilise all the machinery fully would be that three units of A, one unit of B and four units of C should exist. Any expansion or contraction would upset this balance. Suppose only nine units of C existed, then the incentive for adding to C is the greater throughput from the whole system. An expansion in B, for example through new technology, would also create the same effect. This interdependence is probably easier to manage where all the capacity is under single ownership. It is a feature of market situations where only a few firms operate for overall capacity to be greater than that physically required. Decisions about the capacity of individual units are made independently and prices can be maintained through the exercise of market power at levels which keep this excess capacity in existence.

Physical Form
The form of physical plant also determines the number of plans available. A large number of different types of physical objects are theoretically available to carry out a single job. If a job is analysed into its constituent processes and for each of these all the various technologies that can be used are drawn up, a comprehensive description of the number of systems available is made. Consider the job of harvesting grain. Assume that there are four processes involved, namely, cutting, threshing, straw collection and grain treatment. Technological means of carrying out each process are

then assigned as follows:

TABLE 4.6
Methods of Harvesting Grain

Process	Techniques
A Cut	Knife A_1, Flail A_2[a]
B Thresh	Stationary B_1, Mobile B_2, None B_3
C Collect straw	Loose C_1, Sheaves C_2, Chopped C_3, Baled C_4, None C_5
D Treat grain	High temperature drying D_1, Low temperature drying D_2, Chemical D_3, Cooling D_4, Airtight D_5, None D_6

[a]This is a theoretical example. Such techniques are not necessarily found in practice at the present time.

There are two ways of carrying out the cutting process. For the process of cutting and threshing (A, B) there are $A_1 B_1 + A_1 B_2 + A_1 B_3 + A_2 B_1 + A_2 B_2 + A_2 B_3 = (2 \times 3)$ 6 feasible subsystems. Increasing the scope of the subsystems to include straw collection (C) produces $2 \times 3 \times 5 = 30$ feasible subsystems (A, B, C). Hence, the number of ways of carrying out harvesting, provided all the elements will fit together, is $2 \times 3 \times 5 \times 6 = 180$.

Specification is made easier if the number of alternatives for consideration can be reduced without overlooking the best choice in the process. For example, simplification can be achieved in the grain harvesting illustration by removing technically incompatible systems. Thus, the subsystem $A_1 B_1$ can only be added to the elements C_1, C_2, C_3 since the elements C_4 and C_5 are dependent on B_2 being present. Similarly, for each of the other subsystems there are several technically inconsistent means of straw collection as shown below in Table 4.7. In total there are fourteen feasible subsystems of A, B, C. Of these combinations ten produce threshed grain which can be treated in five different ways. Hence, as none of these methods conflict, there are in all fifty-four possible systems A, B, C, D produced from the fourteen A, B, C subsystems.

Competition
At a general level, agriculture has to compete for resources such as

TABLE 4.7
Feasible Systems for Harvesting Grain

Process	Technique						Number of subsystems
A Cut	Knife	Flail	Knife	Flail	Knife	Flail	
B Thresh	Stationary	Mobile	Mobile	Stationary	None	None	
	1	+ 1	+ 1	+ 1	+ 1	+ 1	= 6
C Straw collection	Loose Sheaves Chopped	Chopped None	Loose Chopped Baled None	Chopped	Loose Chopped Baled	Chopped	
	3	+ 2	+ 4	+ 1	+ 3	+ 1	= 14
D Grain treatment	High temperature drying Low temperature drying Chemical Cooled Airtight	—	—	—	None	None	
Total number of systems	15	+ 10	+ 20	+ 5	+ 3	+ 1	= 54

labour and capital with other industries so that the price other industries can afford to pay, directly affects the costs and availability of resources to agriculture. Farmers and small traders and even regions and countries in large international markets are unlikely to be able to force a change in the prices they receive for their products. This is because their individual output is such a small part of the total that it has no significant effect on total supply. However, in situations where large-scale concerns dominate a market, as for example in agricultural supply and processing industries, the pricing policy and services offered are dependent on the actions, both actual and potential, of their competitors.[9] Suppose only two firms, A and B, control a market and that the gains of A are the losses or costs of B. Then as shown in Table 4.8, what each firm does depends on the other's action. In Case I firm A would try initially to pursue a low price policy. However, firm B by also charging a low price would reduce its losses from £70 to £50. In this case, firm A would then charge high prices so as to receive gains of £60. This action would encourage firm B to follow suit with a high priced regime. There is no equilibrium in this case. The result could be that both firms would pursue a mixed strategy taking great care not to reveal their intentions.

TABLE 4.8
Interdependence of Competitive Strategies of a Two Firm Market[a]

			Firm B	
			High price	Low price
			Gains and losses	
Case I				
	Firm A	High Price	£50	£60
		Low Price	£70	£50
Case II				
	Firm A	High Price	£50	£10
		Low Price	£70	£50

[a]In this 'zero sum' situation the gains to Firm A are losses to Firm B.

Case II shows a situation where both firms would stick to a low price policy which would result in a gain to firm A of £50 and a loss to firm B of the same amount. Any other combination would not be stable.

FINDING THE BEST SOLUTION

Within the limits of available resources activities can assume a wide range of levels. What is the best level? Many algorithms exist for solving such problems but their application depends upon being able to construct a model in an appropriate form. For example, calculus can be used if response can be assumed to be continuous.[10] Linear programming can be employed provided the assumptions of linearity, independence, divisibility and additivity can be met. More complicated algorithms can cope with dynamic elements. Risk can be dealt with in the Monte Carlo method by generating a large number of random activity levels and sorting them out according to the degree to which they satisfy the objectives to be achieved. Cruder descriptive models do not find a best solution but simply trace out through simulated time the consequences of various planning decisions. The planner can use such models to test out different situations and so find better solutions although not necessarily the best.

The practical problem which constrains the application of search procedures is whether or not sophistication can be justified. Costs of staff training, data requirements, model construction and computation can be high especially for one-off situations. The benefits of better planning are uncertain because the future is not known and actions in many situations can be modified in the light of experience. Accuracy beyond that of the information available is not attainable. The search process, however, does enable the planner to pick out the important influences on an objective.

Strategy
Strategy or policy is to do with major long-term changes within an organisation and does not lend itself to simple search pro-

cedures.[11] This is because strategy is associated with greater uncertainty and a consideration of time. Less information and more judgement is required in designing decisions of a strategic nature. Strategy provides the framework of production capacity within an organisation and as such it has several features. First of all it wears out slowly and any change requires a large amount of resources in relation to an organisation's total assets. It is a vital part of any institution since it determines to a large extent the ability to compete. It also simplifies day to day management since tactical decisions can be made by reference to strategy. A weakness of many organisations is that no strategic plan has been worked out simply because of day to day pressure of events. No amount of detailed short-term planning can rectify this weakness.

There are two elements in the strategic planning process. Firstly, it involves a thorough analysis of the environment within which an organisation is operating. Secondly, it requires a detailed analysis of the internal strengths and weaknesses of the same organisation. The choice of activity will then depend upon a matching of strengths with opportunities in the environment. It is pointless, for example, for a firm to willingly compete with other firms which have inherent advantages over them.

One of the indicators of strength is the financial situation of institutions as measured by the liquidity of a concern and its degree of borrowing. In either case the balance between liquid and fixed assets and borrowed and self-financed assets is a question of judgement. In a favourable profit situation the higher the gearing or the more money that is borrowed, the greater the profit potential of owned assets. However, the reverse occurs if unfavourable events take place. A hypothetical situation is shown in Table 4.9.

Liquid funds have on their own only a limited earning potential. It is necessary to balance them up with fixed assets which provide the capital structure for a business. A common error in management is to convert short-term or liquid capital into long-term assets in an effort to expand too quickly thus starving the business of cash which can result in a loss of confidence in the organisation by its customers.

TABLE 4.9
An Illustration of Capital Gearing

	High gearing £	Low gearing £
Case I		
Total assets	10 000	10 000
Funds borrowed	5 000	1 000
Net worth	5 000	9 000
Profits before interest	2 000	2 000
Interest paid at 10%	500	100
Return after interest on net worth	$\dfrac{1\,500 \times 100}{5\,000} = 30\%$	$\dfrac{1\,900 \times 100}{9\,000} = 21\%$
Case II		
Profits before interest fall to	750	750
Interest paid	500	100
Return after interest on net worth	$\dfrac{250 \times 100}{5\,000} = 5\%$	$\dfrac{650 \times 100}{9\,000} = 7\cdot2\%$

The people of an organisation, their attitude, abilities and potential are by far the most important determinant of an institution's future success. It is impossible to express the value of people in any formal and systematic way. Management can be simplified in many situations to a maxim of choosing the right people and allocating them to the right kind of job. One of the determinants of performance is the degree of accumulated experience. It takes quite a long time to gain experience and learn how to carry out the functions of agriculture well. This is especially important since care and attention at all stages in the productive process make all the difference. The age structure of a work force in terms of long-term potential determines the flexibility of the problems of handing on responsibility and allowing new ideas and youthful enthusiasm to flourish.

Another strength of the business is its simplicity of organisation. The less links in the chain of command, the less likelihood of things going wrong. Further, the simpler an organisation the easier it is to ensure standby arrangements such as emergency procedures in the event of breakdowns. The ability to stand the shock of adverse events is also important and one way of measuring this is by the margin made per unit of output. Two different policies can be pursued. First, a low margin/high output strategy where a small change in price can reduce the margin altogether which adds to the risk of such a policy. Provided a margin is present, then, combined with high output, profit is assured especially if prices are kept low. The other alternative is the high margin/low output policy whereby prices can move relatively quite a long way before the margin is cancelled out altogether. However, competition is likely to be keen and profits limited by the small size of turnover.

The concentration of decision making is also an important strategic matter. If only a few individuals, or a few institutions, make decisions there may be gains in terms of a reduction in the time taken for action to take place. The problems of coordination will also be less, but it is also possible to increase the chances of large errors which can prove fatal.

Another area where an internal audit should focus is on the use of time within organisations. For example, the existence of seasonal and cyclical price movements means that the timing of sales can be crucial. Delays tend to accumulate and augment each other so that if a crop is sown late it will tend to be harvested and sold late. Once an organisation gets behind then it takes a greater amount of effort to catch up.

An important asset in agricultural organisations is contacts. This is especially important in international trading where procedures are complex and where a knowledge of change is confined to a few individuals. Mutual trust builds up between individuals irrespective of their organisations. When large quantities of money are at stake this is especially important particularly when the costs of disputes and delayed payments and deliveries are high.

The Search Process
The search process consists of adjusting activity and resource levels in a direction that increases the value of the objective. This involves small successive substitutions between inputs and be-

tween products which both releases and uses up resources. Further changes are made until no rearrangement can be found which improves on the value of the overall plan. Once found the best solution can act as a reference point for assessing further changes in the plan such as the provision of more resources. The best plans invariably show up some resources that are not fully used so that adding more of these resources will not increase satisfaction. Activities that are not chosen will also be highlighted while the cost of binding limitations and the effect of their relaxation can be illustrated.

Partial Budgets

Partial budgets are the simplest and most easily applied search method. The approach focuses only on changes in those factors which are assumed to be important. The extra costs of such changes including benefits forgone are compared with the extra benefits including costs saved.

Consider the data on the response of barley yield to nitrogen applications shown in Table 4.10.

TABLE 4.10
Response of Barley Yield to Increases in Nitrogen Application[a]

	Nitrogen (kg/ha)					
	0	40	70	100	130	160
Yield of barley (kg/ha)	1 316	2 322	2 897	3 049	2 847	2 545
Yield increment		1 006	575	152	-202	-302

[a]Price of nitrogen = 40p/kg; price of barley = 10p/kg

Obviously the manager would not consider nitrogen applications above 100 kg/ha since yields decline after this point. If he were to compare the value of the response from applying 100 kg of nitrogen per hectare rather then 70 kg the extra benefits would be the yield response 152 kg multiplied by the price of barley 10p = £15·20. The extra cost of nitrogen would be 30 kg × 40p which is £12. The extra benefits exceed the extra costs so the high rate of fertiliser use would be recommended.

The above example serves to illustrate several of the assumptions of the partial budgeting procedure. Firstly, other influences on yield are assumed to be constant. Secondly, only discrete changes are considered, in this case the difference between 70 and 100 kg of nitrogen per hectare. A comparison between 40 and 100 kg would produce a more optimistic outcome. Thirdly, a positive margin is taken as the criterion for action irrespective of the potentially greater margins that might be earned in some other activity. In the above example, expenditure of £12 could well earn more than £15·20. A partial budget is essentially crude and unless it is intelligently applied the best solution can easily be overlooked. The returns from the 71st kg of nitrogen are likely to be higher than from expenditure on the 100th but the roughness of the partial budgeting approach does not allow such precision. This could be achieved if a quadratic function were fitted to the data and the optimum found using calculus. In this case two quantities determine the best level of application; firstly, the rate of response in yield to extra nitrogen; secondly, the ratio of the price of nitrogen to the price of barley.

A partial budget can also be applied to other types of problem including the best level of competing activities. If activity A is to be increased at the expense of activity B a partial budget would compare the extra revenue from A together with the cost savings from not producing B. The costs would include the lost revenue from B and the extra costs of producing more A.

The substitution between resources can also be explored by comparing the respective costs of achieving a result in different ways. Animals can be fed more concentrates and less roughage. Jobs can be done with less labour and more machinery. Crops can be produced with more land and less fertiliser and so on. A source of confusion in this type of budget is the way in which capital expenditure is treated. Normally capital is converted into an equivalent annual cost which includes both depreciation to cover the loss in value of capital goods and the interest paid on outstanding capital. Another way of dealing with capital is to compare streams of costs and returns over the anticipated life of projects and then to compare the equivalent present value of these streams for different alternatives as previously described in Table 4.2. These concepts can be further illustrated in an interesting way by considering the problem of the best time to replace assets.

Replacement[12]

Consider a machine which costs £12 000 new and which is expected to have annual operating costs and re-sale values as shown in Table 4.11.

TABLE 4.11

Operating Costs and Resale Values of a Machine According to Its Age

Year	Operating costs £	Resale value at year end £
1	105	—
2	188	800
3	262	700
4	370	600
5	310	550
6	356	500

A short cut method of finding the annual cost of a machine is to add together the mean operating costs, the mean drop in value and the interest on the average amount of capital involved. Thus if a machine were to be kept for four years the annual cost assuming a 10% interest rate becomes.

Mean operating costs £	Mean drop in value	Interest on average capital
£105		
£188	(£1 200 − £600)/4	(£1 200 + £600) × 0·1
£262		
£370		
£925/4 = £231	£150	£90

Annual cost = £231 + £150 + £90 = £471

The exercise can be repeated for different machine lives and the life which has the lowest cost found by inspection (Table 4.12). The results at a 10% interest rate show that a two-year life gives the lowest annual cost whereas when capital is more expensive at a 20% interest rate the optimum life is extended to three years.

TABLE 4.12
Annual Costs at Different Interest Rates for Given Machine Lives

Life in years	Equivalent annual cost at 10%	Equivalent annual cost at 20%
2	447[a]	550
3	451	545[a]
4	469	568
5	463	566
6	462	580

[a]Denotes minimum cost

Instead of calculating equivalent annual costs on an *a priori* basis a more realistic decision problem is whether or not to keep an existing machine for one more year.

Suppose a three-year-old machine is already owned and that capital for a new machine could earn a 15% return elsewhere. The stream of costs for the two alternatives of replacing now or in one year's time assuming a four-year life for the replacement machine are as in Table 4.13.

TABLE 4.13
A Comparison of Net Cash Flows from Delaying Machine Replacement

	Replace now	£		Replace next year	£
Year			Year		
	New price £1 200, less resale value £700	−500	1	Operating cost for old machine + New price £1200 less resale value £600	−970
1	Operating cost	−105	2	Operating cost	−105
2	Operating cost	−188	3	Operating cost	−188
3	Operating cost	−262	4	Operating cost	−262
4	Operating cost + resale value (£600)	+230	5	Operating cost + resale value (£600)	+230
Equivalent annual cost =		−270	Equivalent annual cost =		−308

These streams using discounting procedures can be converted into equivalent annual costs so as to ensure comparability. The Net Present Value criterion is not valid since services are provided in the delayed replacement option for a longer time than the replace now alternative. The replace now option under the assumptions made is the cheapest.

It is important in any partial budget to take into account all of the factors that change including taxation and interest charges. There comes a point, however, when in order to estimate the collateral effects of the new plan it is necessary to take a more comprehensive view of the system. For example, in order to estimate the effect of a change on taxable income, the incremental effects would need to be assessed in terms of the whole business with and without the proposal.

Linear Programming[13]
The main drawback to partial budgeting, that of the arbitrary nature of the comparisons, can be overcome by using more systematic search procedures such as linear programming. The application of this technique depends on the feasibility of expressing the planning problem as a set of linear equations which can be solved mathematically. A linear program consists of a set of activities, say X_1, X_2, X_3, X_4. Each unit of these activities will have a pay-off, for example as follows:

Activity	Pay-off per unit
X_1	100
X_2	80
X_3	240
X_4	5

If the aim is to maximise profits then the linear programming algorithm will find the highest value of the total pay-off Z where $Z = 100X_1 + 80X_2 + 24X_3 + 5X_4$ and X_1, X_2, X_3 and X_4 are the amounts of each activity.

If the problem is couched in terms of cost minimisation the aim will be to find activity levels which correspond to the least total cost. However, in optimising Z the plan must simultaneously

satisfy resource use relationships. Each activity will require resources and these can be represented by an input/output matrix as
shown for a hypothetical case below:

Resource	X_1	X_2	X_3	X_4
A	1	1	1	0.1
B	0.5	3	4	0.2
C	1	-1	0	0
D	1	0	0	0

Thus, 1 unit of activity X_1 requires 1 unit of resource A, 0·5
units of resource B and so on.

X_2 in a similar way also requires resources A and B but actually
supplies a unit of resource C.

The final piece of required information is the amount of resources available. Assume in this example that there are 200 units
of A available, 200 of B, 70 of C and 20 of D. The problem then is
to solve the following set of simultaneous equations so as to
maximise Z subject to the condition that activity levels cannot be
negative, as follows:

$$200 \geq 1X_1 + 1X_2 + 1X_3 + 0·1X_4$$
$$200 \geq 0·5X_1 + 3X_2 + 4X_3 + 0·2X_4$$
$$70 \geq 1X_1 - 1X_2$$
$$20 \geq 1X_3$$

The above equations describe all feasible combinations of the
various activities within the resource limits so there is no danger
of overlooking the best solution.

Linear programming is well understood and it has been applied
in many different ways to a large number of problems often
involving the construction of an extremely large number of equations. Variants of the technique enable the method to cope with
time, risk, indivisible or lumpy resources and interrelationships
between activities and constraints. It is used commercially in the
blending of feedstuffs and fertilisers, in enterprise planning, in the
design of storage and transport distribution systems and as an aid
to regional planning. The common feature of all these applications
is that they involve the allocation of scarce resources to activities.

Planning a region's agriculture is conceptually no different from planning an enterprise mix on a farm. Deciding on the least cost ingredients in a feed ration is similar in many respects to finding the cheapest way of transporting a set of products from different sources of supply to a variety of user points.

The development of linear programming was a great advance in planning methodology. Its widespread application was based on the importance of the allocation problem combined with the low cost of constructing and solving linear models. The method does depend, however, on assumptions such as linearity, independence and divisibility of activities. The technical relationships are usually taken to be both deterministic and static while the solution is implicitly a stationary one. The absolute values of the results for all these reasons may not be taken literally by management but they do help to clarify the important determinants of success. In many situations it is necessary for management to investigate the behaviour of systems over time and to evaluate risk. One way of doing this is through the use of simulation models.

Simulation

Simulation unlike analytical models simply describe systems. Their main use is to find out the consequences of different proposals over time. There is no guarantee that the best proposal will be found. Simulation models are also a useful way of representing risk situations. Iterative procedures enable estimates to be made of averages and higher moments for combined frequency distributions. These estimates can then be used to find the most advantageous policies.

Simulation of Time

An example of the way the simulation technique can handle time is provided by describing the growth of a cattle herd in a ranching situation. This problem could be handled by analytical models[14] but simulation while less elegant does enable the model builder to make fewer assumptions about the characteristics of the animals. The logic of the development of the herd is described in Fig. 4.6. It is assumed that the numbers change according to mortality,

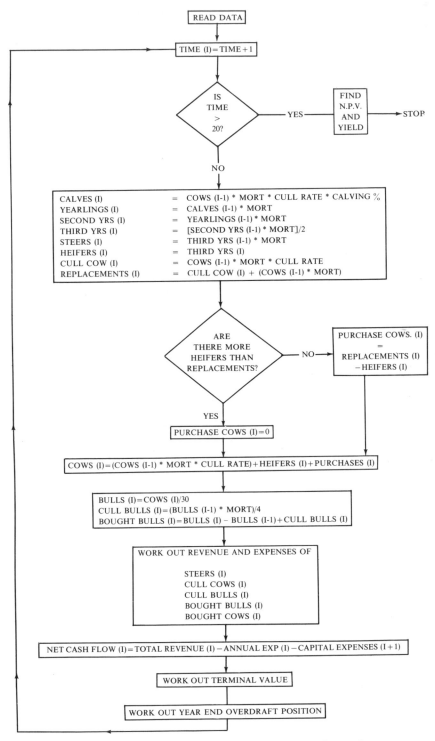

FIG. 4.6. Flow chart of model for forecasting cattle numbers.

culling and calving rates and that an equal number of males and females are born. Heifers are bred when they are three years old and steers are sold at four years of age. In a situation where there are insufficient heifers to match the number of cows that are culled or die then cows are purchased to make up the numbers. Time is advanced in the model on an annual basis. The number of animals generated is used to work out the cash position and the Net Present Value and Internal Rate of Return. A sample output based on the data shown in Table 4.14 is presented in Table 4.15. Changes in the output can be generated according to different assumptions made in the plan.

TABLE 4.14
Data for Cattle Ranch Growth Model

Interest = R	=	10%	Cull cow price	=	50
1 – Calf mortality	=	0·92	Cull bull price	=	150
1 – Yearling mortality	=	0·98	Bull buy price	=	200
1 – Two year mortality	=	0·99	Initial cow numbers	=	100
1 – Three year mortality	=	0·99	Initial calf numbers	=	0
1 – Cow mortality	=	0·04	Initial yearling numbers	=	0
Cow cull rate	=	0·125	Initial two-year-olds	=	0
Calving rate	=	0·4	Initial three-year-olds	=	0
Cow/Bull ratio	=	30	Initial bull numbers	=	0
Bull cull rate	=	0·25	Initial bull purchases	=	4
Calf price	=	20	Initial cow purchases	=	100
Yearling price	=	40	Initial cull bulls	=	0
Two year price	=	80	Initial heifers	=	0
Three year price	=	100	Initial cull cows	=	0
Cow sale price	=	100	Initial revenue	=	0
Steer sale price	=	150	Initial cash flow	=	– 23 800

Simulation of Risk

A way of handling risk in simulation models is by combining distributions using random numbers by what is known as the Monte Carlo technique. Consider the following situation:[15]

A large agricultural contractor wants to know how many combine harvesters he should operate. Because of the weather, he is faced with variable demands for his services, changing operating rates and different numbers of available working days. Further, in a wet year he can expect high demand, lower operating rates and

TABLE 4.15
Sample Output from Program to Generate Growth of an African Cattle Ranch[a]

Year	Cows	Calves	1 year	2 year	3 year	Cullcow	Steers	Heifers	Purcow	Net Cash Flow	Overdraft
0	100	0	0	0	0	0	0	0	100	−23 800	−23 800
1	100	35·14	0	0	0	8·16	0	0	12·16	— 722·67	−27 502·67
2	100	35·14	32·33	0	0	8·16	0	0	12·16	— 848·00	−31 700·93
3	100	35·14	32·33	31·68	0	8·16	0	0	12·16	— 848·00	−36 319·03
4	103·52	35·14	32·33	31·68	15·68	8·16	0	15·68	0	344·53	−40 206·40
5	106·61	36·37	32·33	31·68	15·68	8·45	15·52	15·68	0	2 688·95	−42 138·09
6	109·33	37·46	31·46	31·68	15·68	8·70	15·52	15·68	0	2 702·84	−44 249·05
20	152·65	52·36	47·03	45·00	21·74	12·16	21·02	21·74	0	3 676·70	−99 616·77

[a] Net Present Value at 10% discount rate = −11 244; Yield = 5·75%

fewer days available than in a dry year. How can he take these factors into account? Let us assume he can provide us with the following information:

There are six chances in 10 of having a wet year and four chances in 10 of a dry year.

The chances of particular levels of demand, work rates and available days, in wet and dry years are shown in Table 4.16.

It is necessary to express these distributions in cumulative form

TABLE 4.16
Information Available to Contractor

Wet year		Dry year	
Demand		Demand	
Hectares	Chances out of 10	Hectares	Chances out of 10
700	4	600	1
600	3	500	2
500	2	400	3
400	1	300	4
Rates of work		Rates of work	
Hectares/day	Chances out of 10	Hectares/day	Chances out of 10
5	3	5	1
6	3	6	1
7	2	7	2
8	1	8	3
9	1	9	3
Days available		Days available	
Days	Chances out of 100	Days	Chances out of 100
22	8	22	15
21	12	21	15
20	15	20	14
19	14	19	14
18	11	18	13
17	10	17	13
16	10	16	8
15	8	15	4
14	7	14	4
13	5	13	0

in order to combine them. A level of demand, rates of work and number of available days can be simulated, by chance, using a random number according to a matching of the value of the random number with the range of the cumulative distribution in which a particular value falls (Table 4.17). A sequence of 'n', where

TABLE 4.17
Cumulative Distributions

Wet year		Dry year	
	Demand		
Range of random number in which value chosen	*Value*	*Range of random number in which value chosen*	*Value*
0– 9	400	0– 9	600
10–29	500	10–29	500
30–59	600	30–59	400
60–99	700	60–99	300
	Rates of work		
0– 9	9	0– 9	5
10–19	8	10–19	6
20–39	7	20–39	7
40–69	6	40–69	8
70–99	5	70–99	9
	Days available		
0– 7	22	0–14	22
8–19	21	15–29	21
20–34	20	30–43	20
35–48	19	44–57	19
49–59	18	58–70	18
60–69	17	71–83	17
70–79	16	84–90	16
80–87	15	91–94	15
88–94	14	95–99	14
		—	13

'n' is a large number, can be simulated by repeating the procedure 'n' times (Table 4.18).

For the purposes of this illustration 25 'seasons' have been simulated using random number tables and the number of com-

TABLE 4.18
Results of Simulation Over 25 Seasons

Season	Demand	Rate of work	Available days	Combines required (rounded up)
1	Wet	700	6	21	6
2	Wet	700	7	20	5
3	Wet	500	6	22	4
4	Wet	700	7	22	5
5	Dry	600	7	16	6
6	Wet	600	6	20	5
7	Wet	500	9	13	5
8	Dry	600	8	20	4
9	Wet	400	8	19	3
10	Wet	500	7	17	5
11	Wet	500	8	17	4
12	Wet	700	5	15	10
13	Dry	300	9	19	2
14	Wet	600	7	14	7
15	Wet	700	7	14	8
16	Wet	500	5	20	5
17	Dry	300	5	18	4
18	Dry	300	7	17	3
19	Dry	300	7	22	2
20	Dry	300	5	16	4
21	Dry	500	6	20	5
22	Dry	300	8	21	2
23	Wet	700	8	17	6
24	Dry	500	9	18	4
25	Wet	600	5	20	6

bines required per season rounded up to whole numbers as is shown in Table 4.19.

The final decision as to how many combines to operate will depend on the costs of holding extra machines as against the costs of failing to fulfil demand in some years. The simulation by combining the distributions provides the contractor with a measure of the risk he is taking by operating different numbers of machines. (The combined distribution is an important element in techniques of investment appraisal for use in risky situations.[16,17])

TABLE 4.19
Frequency with which a Number of Combines are Required

	No. of combines								
	2	3	4	5	6	7	8	9	10
Frequency	3	2	6	7	4	1	1	0	1

He would probably not operate the modal number of machines (5) in that if he did so he would not finish harvest in seven years out of 25 or 28% of the time.

ACTION

The selection of a plan for action depends on a final check on its feasibility. The most common causes for hesitation are financial restrictions and the risk involved. In many situations it is not until a plan is proposed in detail that the true nature of a decision maker's objectives are actually discovered. Further the examination of a plan often reveals that the wrong problem has been specified. In either situation it will be necessary to go through the planning sequence again. It is of course a common experience for managers to be at a loss about what to do especially with human problems.

Planning is not a once and for all activity. Before implementing a plan it is wise to consider the proposals in terms of what might go wrong and to devise contingency plans in case the worst happens. The necessity to do this will depend on the likelihood of adverse situations occurring and the extent to which errors in decision making can be reversed. Usually some means of compensation can be found especially if plans are adopted which have a built-in second chance. Things are more likely to go wrong when plans involve a great deal of change. New, complex and unfamiliar activities are especially risky. Tight restrictions in terms of time, capacity and finance also increase the danger of failure. Furthermore, difficulties in the implementation of plans are more likely to be experienced if responsibility for them cannot be easily assigned. This would apply to plans which influence more than one management unit.

REFERENCES

1. LINDLEY, D. (1971). *Making decisions*, Wiley Interscience, New York, Chapter 7.
2. ROUMASSET, J. A. (1976). *Rice and risk. Decision making among low-income farmers*, Elsevier North-Holland Publishing Company, Amsterdam, Chapter 2.
3. WRIGHT, R. (1964). *Investment decision in industry*, Chapman and Hall, London.
4. SPEDDING, C. R. W. (1979). *An introduction to agricultural systems*, Applied Science Publishers, Ltd., London.
5. GILES, A. K. and STANSFIELD, J. M. (1980). *The farmer as manager*, George Allen and Unwin, London.
6. MISHAN, E. J. (1972). *Cost benefit analysis: an informal introduction*, George Allen and Unwin, London.
7. BIERMAN, II. and SMIDT, S. (1966). *The capital budgeting decision—economic analysis and financing of investment projects*, (2nd Ed.). Collier-Macmillan Ltd., West Drayton, Middlesex.
8. HALTER, A. N. and DEAN, G. W. (1971). *Decisions under uncertainty with research applications*, South-Western Publishing Co.
9. WILLIAMS, J. D. (1966). *The compleat strategyst*, McGraw-Hill Book Company, Maidenhead.
10. DILLON, J. (1968). *The analysis of response in crop and livestock production*, Pergamon Press, Oxford.
11. WHITTAKER, J. B. (1978). *Strategic planning in a rapidly changing environment*, Lexington Books, D.C. Heath and Company.
12. WRIGHT, F. K. (1960). *The replacement problem*, Supplement to the *Australian Accountant*, September 1960, entitled: *The economics of capital expenditure*, by K. A. MIDDLETON.
13. HEADY, F. O. and CANDLER, W. (1958). *Linear programming methods*, Iowa State University Press, Ames, Iowa.
14. WILLIAMSON, M. (1974). *The analysis of biological populations*, Edward Arnold, London.
15. NIELSEN, A. H. (1968). *Dimensioning of resources. An example of the use of queue theory*, Reprint from *Journal of Agriculturalists*

(*Ugeskrift for Agronomer*), Economic Institute, Royal Veterinary and Agricultural University, Copenhagen, pp. 735–741.

16. REUTLINGER, S. (1970). *Techniques for project appraisal under uncertainty*, World Bank Staff Occasional Papers Number 10.

17. POULIQUEN, L. Y. (1970). *Risk analysis in project appraisal*, World Bank Staff Occasional Papers Number 11.

CHAPTER 5

Control

Agricultural systems if left to their own devices decline and disintegrate. They go out of control. Animals stray, crops fail, pests proliferate, buildings and machinery fall into disrepair, costs rise, output falls and markets are lost. An essential task of management is to confine system performance within preferred limits. These limits may be expressed in terms of means, variance or skewness either separately or together. Businesses for example may seek to attain a long run average level of profitability. However, in order to survive they must also achieve a given level of profit every year.

The simplest procedures for controlling systems are very much a matter of common sense. Thinking ahead, so that preparations are made in good time, saves last minute rushes and reduces the chances of mistakes. Finishing off one job where possible before beginning another ensures orderliness in management. Making sure that jobs are done well minimises the chances of failure.

Control is made easier if the variety of a system can be reduced. One way of achieving this is to classify objects into homogeneous groups so that there is no need to observe all the components of a system but only those within critical groups. Economy can also be achieved by sampling procedures, the degree of accuracy required determining the type of sample and its size.

The concept of system stability can be thought of as the

association between the desired system output and actual performance. Changes in both the environment and the system itself will produce deviations from the preferred state which will vary in frequency, magnitude and duration.

Three general ways of controlling a system's output have been described by Lange.[1] The first method consists of making up the difference between the preferred state and the actual state by equalising the deviation. For example, if an animal is sick, treatment is administered which overcomes the symptoms without necessarily removing the cause.

A second method is to compensate for disturbances. Thus if the relationship between predisposing environmental changes and illness is known, sickness can be prevented by prophylactic measures such as hygiene, vaccination and the removal of intermediate hosts.

The third measure is to eliminate disturbances. This is not always feasible. It is not possible to change the weather for example but artificial environments can be created. In the case of sickness a disease can be eradicated or the spread of the causal agent restricted by quarantine arrangements.

EQUALISATION OF DEVIATIONS

The process of equalising deviations in system performance involves the comparison of the actual output state with some norm or standard. When some critical deviation occurs action is taken to correct this. Management therefore has to monitor the state of the system, set standards and make decisions about what action should be taken and when. The whole process can be thought of as a closed system, that is there are links or relationships between the level of output and the level of input.[2,3] Achieving control by this means is known as feedback as represented in Fig. 5.1. The feedback mechanism can either amplify or damp down the output through its effect on the input as dictated by the state of the output. As the system goes out of control so corrective action is initiated.

The control of agricultural systems by the use of feedback mechanisms is often done on a trial and error basis. In some cases,

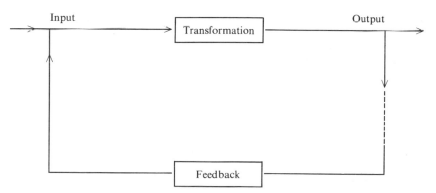

FIG. 5.1. The logic of feedback control.

feedback mechanisms are not used, as for example in project planning when no evaluation is carried out. In other cases, the setting of appropriate standards is made difficult by inaccuracies and lags in the information received about the state of system output. This adds to the difficulty of deciding on appropriate forms of corrective action.

Setting Standards

Consider a merchant who is faced with uncertainty about the demand pattern for a particular range of goods. The holding of stocks will enable him to supply his customers in the event of an unusually high level of demand or if his normal supplies are delayed. The merchant is thereby able to even out the effect of fluctuations in both demand and supply. His problem is to decide on the quantity and type of goods to store. This can only be solved by a consideration of the costs and benefits of holding stocks in conjunction with estimates of the likelihood of particular states occurring.[4] If he stores too many items, while he may satisfy most of his customers most of the time his costs may exceed the benefits derived. If he holds too few items he may lose a significant number of customers. The costs will include charges for capital represented by the stocks and the storage space they occupy. These will need to be balanced against the costs of not being able to meet demand if stores are not kept, weighted by the chances of such events occurring. The level of costs will vary. Large expensive items will cost a lot more to store than small inexpensive ones.

Some goods will be purchased regularly by long-standing customers, others will be bought infrequently by non-regular clients. The costs of potential lost custom will therefore vary with the item.

The merchant will be able to improve his judgement about stock levels if, for example, a seasonal pattern of demand is discernible. In this case the desired stock levels will be a function of time. He may also be able to anticipate changes in demand and supply by an understanding of the reasons why changes take place. Demand for example may be influenced by the weather, technological change or changes in the structure of the industry. The rewards for correct anticipation of the state of the market can be high. The merchant will also improve his decisions about stock levels by learning from past experience. He will adapt to change according to the rewards and penalties received from his past decisions.

The Need for Feedback
The effectiveness of the aggregate behaviour of the large number of separate institutions contributing to agricultural output is much improved if there is communication between them. This can be demonstrated in the following example. Suppose a marketing chain consists of four retailers, two wholesalers and one manufacturer (see Fig. 5.2). Assume that each retailer has a weekly demand of two units and keeps stocks equal to demand, i.e. two units. Given stable conditions the wholesalers will receive weekly orders of four units each, or a total of eight units.

If wholesalers follow the same rule for stockholding as retailers they will keep stocks of four units each. The manufacturer under stable conditions will receive orders of eight units per week.

Assume that there is a temporary increase in demand at the retail level to three units per retailer. If each retailer believes this is a permanent change, in other words he behaves mechanically without the benefit of previous experience or any knowledge of future market behaviour, then he will not only send out orders that reflect the increase in demand[3] but he will also endeavour to

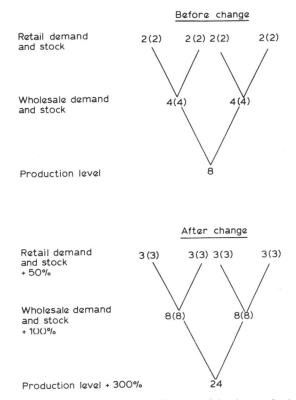

FIG. 5.2. The effect of changes in retail demand in the marketing system.

add one unit to his stock levels. The wholesalers will in consequence receive orders of four units from each retailer or double
the normal order level. Similarly this level of orders will be
amplified yet further at the manufacturer's level. He will receive
orders three times the normal level.

The opposite effect would occur the next week if retail demand
levels returned to normal. The manufacturer would receive no
orders at all as retailers and wholesalers met demand out of
stocks.

The amplification of change in this particular system could be
moderated if all concerned had more information about the true
state of the market. Even if the change was thought to be

permanent a more gradual adjustment of stock levels could be achieved if stocks were used primarily to even out the effect of changes. If coordination was achieved, it would be possible for example for the manufacturer to add one unit of production to his weekly schedule for four weeks to restore the equilibrium. This coordination might be done by direct communication of sales volumes, stock levels, manufacturing capacity or surveys of buyers' intentions. Another way would be to raise the prices of goods as demand increased and to reduce them as demand fell.

The need for feedback is emphasised when it is appreciated that changes in output reflect both adjustments in demand and stock level. Suppose total consumption remains constant at eight units per week. Then, as shown in Table 5.1, the change in output reflects the rate of reduction in stock levels. When stocks fall in week 2, output plummets, but rises again in week 4 even though stocks are still being cut.

TABLE 5.1
Changes in Output as a Function of Changing Stock Levels

	Week 1	Week 2	Week 3	Week 4
A. Manufacturing output	10	6	6	7
B. Consumption	8	8	8	8
C. Change in stocks A − B	+2	− 2	−2	− 1
D. Percentage change in output		−40	0	+16

The above concepts have value for controlling the constituent parts of agriculture. They should be especially interesting to firms which supply inputs to agriculture and for those involved in distribution. The principles can also help in understanding the contribution of agriculture to economic development. Changes in agricultural output can be multiplied up in the regional or national economy. The growth of agriculture will add to the demand for services from the rest of the economy. Firms supplying these resources will in turn also need to invest in new facilities and employ more staff. The size of the multiplier effect will depend on the extent to which output can be increased from existing capacity and on the degree to which local as opposed to imported goods and services can meet the demand.

Lags in Feedback Control Systems

It often takes a considerable amount of time to recognise and measure the actual state of a system's output, with the result that the application of corrective measures is imprecise and can even be disruptive. Attempts, for example, to regulate markets using buffer stocks could increase the amplitude of price variation if the timing of sales and purchases were misjudged. The manager of a buffer stock can never be sure that he is acting wisely.

The long delay in responding to market changes in the case of breeding livestock or tree crops can also produce production cycles, the most notorious of which is the pig cycle. High prices attract new producers to the industry but by the time their plans are on stream total production will have increased. The level of current prices is an inaccurate indication of the true state of the market. This is especially so when it is realised that in the expansion phase of the pig cycle the proportion of younger animals which have lower reproductive rates will be higher than in the contraction phase.

A further reason for the mismatching of production and consumption in agriculture can be called the ratchet effect. A period of high prices encourages an increase in the productive capacity of the industry. This investment, in fixed equipment, land improvement, buildings and machinery, can take several years to install. It is associated with favourable expectations for the future which can persist for some time even though markets have changed. Confidence for example may be based on historic accounts. In so far as statements of profit tend to be produced some time after the end of a financial year they are highly unreliable guides to future profitability. Once fixed investments have been made they become trapped in the industry because they are not easily transferred to other uses. The result is that they continue to be used for a long time even though the prices of agricultural output fall substantially.

The effects of lags can be illustrated by the problems faced by a person wishing to take a warm shower where the temperature of the water is controlled by a mixer tap. Feedback is arranged in the closed system shown in Fig. 5.3 by adjustments to the tap according to the warmth of the water hitting the person's body.

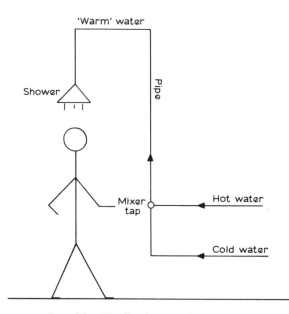

FIG. 5.3.   Feedback control in a shower.

The longer the pipe between the tap and the shower the more difficult it will be to set the tap so as to produce the right temperature. This is because the control process alters the temperature at the mixer tap and not at the shower outlet. The controller has to guess how long it takes for his actions to take effect. Even worse problems arise if there are fluctuations in the temperature of the supplies of hot and cold water. Water which is too hot or too cold may result in the system breaking down altogether, maybe permanently.

COMPENSATION FOR DISTURBANCES

Controlling systems according to this principle depends upon the availability of knowledge in the form of quantified relationships between the source of the disturbance and the desired system state. The operational problems of a shower could be overcome if a sensor within the mixer could be designed which would measure and alter the respective flows of hot and cold water to give the desired temperature of the mix.

There are many examples of the compensatory control principle in agriculture. Breeding programmes for plants and animals can be thought of in this way. Knowledge about relationships between the genotype and performance in particular environments acts as a guide to selection and crossing programmes.

Disease and pest control can also be made more efficient if knowledge exists about pre-disposing factors. Otherwise control measures such as the application of chemicals may be applied on a blanket basis irrespective of the occurrence of conditions favourable to the disease. Prophylactic measures are usually much less costly than dealing with a disease once it has occurred since in this way damage is avoided.

The principle of compensatory control can also increase the effectiveness of financial management. Comparisons are made between actual and desired levels of financial performance in most organisations on the assumption that a closed loop system exists. The standards can be based on historical records but as already pointed out these can be less than satisfactory measures of desired performance. Managers are therefore asked to produce budgets of future receipts and expenses which are then used to assess performance. However, apart from the tendency of managers to under- or over-estimate future prospects according to their objectives, this approach ignores the fact that changes in prices, yields, resource levels as well as windfall gains and losses invalidate the original assumptions used in determining the budget. The realism of target levels of performance can therefore only be retained over time if the effects of changes can be incorporated. Financial performance should ideally be measured in relation to a moving target.

There are usually considerable errors attached to forecasts of the likely effects of disturbances within agriculture. A manager has, therefore, to weight the costs and benefits of his control measures by the chances of them being successful. Inevitably this is a subjective process and actions are made on the basis of hunch and feelings. Those who would influence the control process of agriculture are therefore in the persuasion business. It is not sufficient to develop a reliable compensatory control mechanism. It has to be presented to the industry in such a way that expectations about its effects are well enough understood to change old methods.

ELIMINATION OF DISTURBANCES

External Changes
The effect of external disturbances on a system's output can be avoided by removing the disturbance itself. This presupposes some control over the environment which is something of a paradox. Nevertheless it is true that in many instances through the joint action of many individual institutions the environment can be internalised. A good example of this is where through cooperative activity the members can equalise bargaining power in the market place, they can afford large plant and so exploit economies of scale and run mutual insurance schemes.

Where the main source of disturbance is the activities of other institutions such as suppliers and processors, then the process of vertical integration has the effect of bringing this source of change under control. However, the success of this policy depends on how well the internal functions of the various divisions of an institution can be coordinated. This can be just as difficult a task as dealing with a separate body. The procedures for measuring the contribution of individual divisions to the whole needs to be carefully worked out. For example, the arbitrary allocation of overheads can distort actual costs. A large current deficit may hide the fact that internal transfer payments are wrongly valued and overlook past contributions to profit. Where possible, divisions should be free to trade with outside groups even if internal supplies are available. In this way a meaningful recognition of true prices within the organisation will be retained.

The existence and importance of futures markets is another illustration of how external effects are passed on to those most willing to assume the risks involved. The activities of speculators mean that traders and suppliers of physical commodities can fix their prices ahead of time. They are thus able in turn to offer fixed-price contracts to their suppliers.

Internal Changes

Disturbances can also be dealt with by appropriate system designs. One device is that of spare capacity being carried within the system to cope with emergencies. An animal which is frightened can run away, roll up or take up some other defensive position. Businesses and governments can adopt crash programmes in response to crises. These measures are not usually sustainable. Thus, when a contagious disease breaks out governments impose restrictions on animal movements; overtime is worked during good harvest weather and emergency arrangements for food supplies are made during a famine.

Fail-safe mechanisms in system design are also used to compensate for breakdowns within the system itself. These devices enable management to switch to an alternative without undue strain.

Fail-safe mechanisms are important in sequential operations where the output from one part of a system is the input for a succeeding part. This type of arrangement is a feature of agriculture as discussed in Chapter 2. Other examples include machinery operations, hierarchical chains of command and communication systems.[1]

Suppose we have three machines operating in sequence and each machine works satisfactorily 90% of the time or is broken down 10% of the time. If one machine breaks down the whole system fails. The chances of all three machines working at any one time using the multiplicative probability rule would be $0.9 \times 0.9 \times 0.9$ or 72.9%.

The reliability of the whole system is dramatically improved if an alternative machine at each stage can be employed in the event of breakdowns. The chances of two machines failing at the same time becomes 1% (0.1×0.1) and for the whole system the reliability would rise to $0.99 \times 0.99 \times 0.99$ or 97%.

Reliability can also be improved by reducing the number of links within a sequential process. Thus in the above example, if there are only two machines in a system the reliability with and without duplicates, assuming the same chance of breakdown (10%), becomes 81% and 98%. This principle is often invoked during emergencies where top management assumes direct control of operations. However, such actions are not sustainable since during the emergency other duties are neglected. Fortunately, the same effect can be achieved in more normal circumstances by de-centralising decision making, since the nearer the decision maker is to operations the shorter the chain of command. In consequence, an aim of management in designing reliable systems is to delegate decisions as near as possible to the action; in the case of farmers to workers; in the case of large concerns to divisions; in the case of governments to institutions. The case for decentralisation becomes even more powerful given the increasing difficulty and the cost of duplicating higher levels of management and the fact that centralised management has to rely on long command chains which are unreliable and slow in operation. Decentralisation also harnesses the desire of most people to assume responsibility which in turn improves motivation and care.

The existence and importance of reserves or alternatives in the design of reliable systems can also be recognised in the case of storage.[5] Consider a farmer who can either receive a high price (£100) or a low price (£50) with equal probability (0·5) (Fig. 5.4). His expected price will be £75. If a store is available and he adopts the rule that if the price is low he will store his produce rather than sell, then assuming the probabilities of receiving the same high and low price do not change, the expected value becomes £87·50. This is because the existence of the store gives the farmer a second chance of receiving a high price.

The fail-safe and storage alternatives improve on the reliability of a system by increasing its variety. This is also true where management diversifies its activities since more components with their associated characteristics and relationships are introduced into the system.

The success of diversification is dependent on the extent to

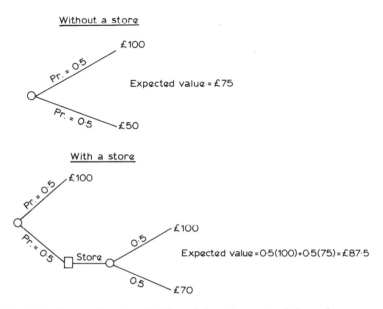

Fig. 5.4.   Improving the reliability of the price received through storage.

which different circumstances produce compensatory effects in separate activities. Consider the hypothetical situation in Table 5.2 where two crops react in different ways to wet and dry seasons.

TABLE 5.2
Pay-offs from Activities with Potential for Successful Diversification

	Wet season	Dry season	Mean pay-off
Crop A	£100	£90	£95
Crop B	£90	£100	£95
$\frac{1}{2}$ A $\frac{1}{2}$ B	£95	£95	£95

If both types of season are equally likely then the average pay-off from specialisation in either crop would be the same, namely £95. However in some years returns would be as low as £90 and in others as high as £100. The variation in returns can be removed entirely in this case by growing a mixture of A and B in equal proportions. The outcome in any season would then always be the same at £95.

Where compensation is less than perfect the proportion of each crop which is grown also determines the success of a diversification policy. Consider the case in Table 5.3 where a negative association between pay-offs still exists but not to the same extent in the different types of season.

TABLE 5.3
Variation in the Degree of Negative Association Between Different Activity Pay-offs

	Wet season	Dry season	Mean pay-off
Crop A	£100	£95	£97·50
Crop B	£90	£100	£95·00
$\frac{1}{10}$ B $\frac{9}{10}$ A	£99	£95·50	£97·25

In this case complete specialisation in A would give the highest average outcome. The worst that could happen if A were grown alone is to have a margin of £95. This lowest value can be raised if some of crop B is grown, but only at a cost. This is because the difference between the respective pay-offs in a wet season is twice as great as it is in a dry season. Thus growing one-tenth B and nine-tenths A reduces the spread of outcomes, but the mean pay-off also falls as compared with growing entirely A.

CONCLUSIONS

The choice of a control mechanism ultimately depends on cost. All controls have a cost, and management has to balance this with the benefits controls bring. These costs include the time and effort required in feedback systems to set standards and to monitor system performance. The tighter the bands of acceptable performance and the more frequently observation is required the greater the precision (and cost) required of a control system.

Research and development costs will be incurred in the design of compensatory mechanisms. Such devices have the appeal of elegance and technical efficiency. However, in the case of disease control, for example, blanket treatments may still be applied if the cost of the treatment is low, the costs of error are high and the detection of predisposing causes is troublesome or subject to doubt in the minds of those who take decisions.

Modifying the environment is usually difficult. However, the reliability of systems built from unreliable parts can attain remarkably high levels if the variety of systems is exploited by producing alternatives and shortening the length of sequential processes. Storage and diversification can be thought of as special cases of the same mechanism. However, in so far as duplicates are required to increase reliability, these will add to cost.

The costs of direct controls foster the search by management for self-regulating systems. Systems can be designed for example which carry out the processes involved in closed loop control automatically. Management may still need to set standards but even this job can be delegated to others provided the right conditions exist for people to do this. Incentive schemes of various kinds, for example, are part of the tool kit of personnel managers. Perhaps a more important task for general management is the removal of disincentives for the achievement of objectives or in more colloquial terms management needs to concentrate on carrots and rather less on sticks.

In the ecosystem self-regulation is the norm. Natural flora and fauna flourish when given an appropriate opportunity but in the main provided natural conditions are maintained environments are not overrun by a single species. This is because of the checks and balances within ecosystems. Resources are limited, predators are common and there is much interdependence between species. The essence of control is thus meaningful interaction of self-determining organisms or competition. Brutal as this may be it works. There is no Ministry of rabbits or weeds. The strong survive or multiply, the weak or unsuited fail or die through competition for resources.

The same mechanisms operate in markets through prices. Prices on a given day clear markets but they also serve to show the degree of imbalance in the market to both producers and consumers. Moreover, prices actually distribute rewards and costs so that they become the means by which those who produce and consume are chosen. Thus higher prices not only indicate to consumers that they will have to pay more to eat but because they have to purchase the higher priced food they cannot with fixed incomes actually buy the same amount. Similarly lower prices for some producers will mean that they cannot meet their expenses and they will cease production. These decisions about who will produce and who will consume have to be made when resources are limited whether in a free enterprise economy or in a centrally planned and administered one. The attraction of the price mechanism is that in theory at least it operates efficiently with minimum interference by management. The price mechanism is not, however, perfect. Conditions need to be designed so as to avoid exploitation. For example too great a concentration on either producers or consumers can result in the creation of monopoly power.

Another weakness of the market system occurs for those commodities whose benefits cannot be confined to those who produce or consume them. The receiver of these goods can obtain them without payment so that in the end if the producer cannot cover his costs they will not be produced. Such so-called public goods have to be provided by some other means than the market. In agriculture public goods include price information, research, education and other physical forms of infrastructure such as roads and communications.

REFERENCES

1. Lange, O. (1970). *Introduction to economic cybernetics*, Pergamon Press, Oxford.
2. Distefano, J. J., Stubberud, A. R. and Williams, J. J. (1967). *Theory and problems of feedback and control systems*, Schaum Publishing Company, McGraw Hill Book Company, Maidenhead.
3. Forrester, J. (1961). *Industrial dynamics*, MIT Press, Cambridge, Mass., 464 pp.

4. HADLEY, G. (1967). *Introduction to probability and statistical decision theory*, Holden Day, Inc., McGraw Hill Book Company, New York.
5. UPTON, M. (1976). *Agricultural production economics and resource use*, Oxford University Press, Oxford.

CHAPTER 6

Recording

The manager can greatly simplify the recording process by having a clear view of why he wants to record facts about his system or events within the environment. The variety of a set of records can potentially be just as great as the variety of the system being measured. The ultimate purpose of records, which can only be historical, is to guide future activity. They are necessary for carrying out the processes of planning, control and forecasting.

All records reflect hypotheses about reality. A set of records designed to prevent fraud, for example, would differ in both detail and design from records to be used for financial control. Unfortunately it is not possible to anticipate all potential management problems so that records designed for one purpose are often the best available information for the solution of different sets of problems.

The costs of recording can be high. Resources such as time and equipment are required to define what is to be measured, to devise appropriate methods of measurement and to design handling, sorting and storing procedures. The attitude towards record keeping is frequently poor especially if the end point or the benefits from the process are not apparent to the recorder. Organising feedback between the recorder and the uses to which his information is put encourages reliability.

Recording errors can be costly to management and it is therefore necessary to check on the consistency of records. One way of doing this is to compare different estimates of the same quantity. For example, in most egg production units egg sales are recorded for financial purposes while laying performance is noted in order to check on health and feeding requirements. The number of eggs sold should equate with the number of eggs laid after allowing for breakages and changes in stocks. The difference between the two measures of the number of eggs in existence gives an indication of the accuracy of the recording process. The same principle is used in the design of experiments where treatments are replicated so that more accurate measurement of their effects can be made. The procedure also produces estimates of the degree of error.

Information can be built up from records through aggregation, by calculation and by transformation. A dairy herd manager, for example, may multiply up milk yields collected at one milking per week in order to estimate herd yields. He may graph the data he has against time and compare his estimates with some standard in order to make him aware of any unusual variation in performance. He may relate his estimated herd yield to climatic variables or use these yields in conjunction with the milk price to construct a financial budget. All this information is built up from the recording of milk yields once per week. Again in the national context aggregate concepts such as the agricultural sector's output and expenditure can be estimated by raising farm sample results.[1] The consistency of the raised sample results can be checked by comparing them with measurements of total sales and purchases of firms which service farmers.

Records can be used to rank performance by individual firms or institutions through comparison with other like bodies operating in similar conditions. Great care is needed in such exercises but an awareness of rank and of progress or decline is helpful to management. Comparison can be made more accurate by adjusting data to take account of differences in the characteristics of the institutions being compared. Particular care over definitions of what is being measured is required such as ensuring that the time periods over which measurements are taken are the same.

A practical recording problem is to identify the subject for measurement in such a way as to be able to keep a track of it through time. This applies for example to individual animals, fields, machines and transactions. Communication between the various parts of agriculture is made feasible if items can be traced backwards and forwards along the production marketing chain. Sampling procedures are often the greatest source of error in analyses of soils and plant material. The problem of sampling the contents of a silage pit given the difficulty of access is an illustration of how difficult the sampling process can be. Comparing the performance of businesses within the agricultural sector is also subject to sampling errors as well as the problems of being unable to compare like with like. This is especially difficult when inter country comparisons of agriculture are made. The choice of exchange rate is a critical decision in this kind of exercise as is checking on the consistency of definitions.

THE RECORDING OF EVENTS

Records of events such as the purchase of land and the details of contracts entered into are most important in management. They can be referred to when things go wrong, which is helpful for all parties in times of stressful change. Events such as dismissal, death, redundancy, eviction or succession can be made much less damaging if they are thought about before they occur by agreeing procedures and writing them down. This is especially so in the case of ownership, inheritance and use of the major capital assets in agriculture, such as land, and in the transfer of the management control of these assets between generations. The normal form of business in agriculture is the sole proprietor so that the life events of individuals have an important part to play in the success of any business. For example any deterioration in personal relationships between father and son, landlord and tenant or between individuals and the representatives of governments and firms can all result in a fall in productivity.

Clear laws and practices can produce a less stressful life for all concerned and reduce burdensome, expensive and protracted litigation. Even simple records such as the details of animal movements can be of invaluable assistance in the early detection and control of disease outbreaks.

Improvements in efficiency can also be gained if trading activities are conducted according to fair procedures and rules. The proper description of food products builds up trust and confidence and reduces costs. Buyers can acquire goods without the need to inspect them provided the description or grade is accurate. Bulk transport of such commodities as grain is possible without the need to keep individual consignments separate. Inspection, weighing and classification in such cases are often done by independent agencies so that all concerned are sufficiently satisfied concerning the impartiality of the recording body. Markets of any kind can only operate where there is confidence in contracts to supply or take delivery of goods. The conduct of markets, including rules for payment, guarantees, penalties for malpractice and appeal procedures, are all vital necessities in the conduct of trade.

Management spends a great deal of time in making sure that recording is accurate and some general rules need to be followed. As mentioned previously, those who do the recording should understand why it is necessary and preferably should be able to see the benefits from their recording activity. Care should be taken in the type of medium on which recording takes place. It is unfair to ask a stockman or a machine operator to write things down while they are carrying out their normal duties. Forms designed by Government Departments should recognise the level of literacy that respondents are likely to comprehend. Language should be simple and clear and questionnaires should not take too long to fill in. Simple instructions should be available. Definitions should be clear and procedures simple and systematic. Records should be permanent and every effort should be made to ensure that they need not be copied or transcribed since errors will inevitably result. Designing record forms so that data are collected in the order in which they are to be analysed can help with this problem.

Finally, there is nothing more frustrating to management than being unable to find a record when it is required. Methods of filing and storing data need to be well worked out. It is much easier to design a filing system if a clear idea of what information will be required is known. This principle applies with equal weight to both manual and computerised data handling procedures.

THE USE OF RECORDS

Some checks on the accuracy of knowledge about systems can be made by using identities. For example, within a given time period, the size of a livestock population can only change by either death or other forms of loss, sales, births and purchases. A check on livestock records can be made by testing out the following equality. The closing stock of animals should equal the opening stock of animals less deaths, less sales, plus births, plus purchases. Similar kinds of checks can be made with financial records, for example, for cash transactions. The closing bank balance should equal the opening bank balance plus receipts less payments made. Stocks of goods can also be checked in a similar way.

The most basic financial statement for any kind of institution is a detailed record of financial transactions and these in small businesses are usually kept in a cash book. Within the cash book, receipts and expenses together with explanatory details are noted in chronological order. In addition, receipts and expenditures may be categorised into trading items, personal items and capital items. Further classification may be made according to the type of activity to which the transaction belongs. This may be by enterprise or by the type of commodity or input that is traded. A simple check on the accuracy of a cash book can be made by comparing the sum of all the individual financial transactions with total receipts and expenses. This in turn can be checked for its completeness by comparing the totals in the cash book with the bank statement.

The entries in a cash book are the basis of accounts. Checks are also made in the construction of accounts according to definitions and the interactions between different types of accounts. To describe these it is necessary to have an understanding of such concepts as profit and the wealth of an individual.[2]

Profit

Profit is a flow concept. It seeks to relate the output in value terms of goods and services produced per unit time with the resources used in the production of these goods and services. Profit is the residual between receipts and expenses. The time period chosen is normally a year but can be of any length. The use of inputs and outputs generated within a given time period is not accurately reflected in the actual financial transactions. A business may have produced output which has not been paid for or which has been added to stocks. Inputs may have been provided out of stocks or payments still be owed to the suppliers. Thus, as shown in Fig. 6.1, it is necessary to make adjustments to the difference between cash receipts and payments for changes over the year in the amounts owed by and to the business and for alterations in the stock levels of both inputs and outputs.

A further difficulty in deriving profit is that for many small firms the separation of business and personal affairs is not clear cut. Most farmers, for example, operate as family units and personal transactions are inextricably bound up with the business. The main input used is the farmer's and his family's labour for which no cash payments are made. In many situations the main product of farming activity is food for the family and again there is no cash transaction. However, it is rare to find a completely subsistence type of economy and most farmers whatever their size do enter into the cash economy. Even large farming businesses still have elements of personal expenditure and receipts within their business affairs. One of the more intractable problems in defining what constitutes a business is for non-nuclear types of family where members make both separate and collective contributions to the whole. Individual members may run separate businesses as well as contributing to a large and loosely defined family enterprise. This same problem arises, paradoxically, with large organisations which are separated out into specialised units which may be jointly run and owned with outside concerns but nevertheless are managed by one central administration.

Profits can be over or under estimated by the valuation of stocks in different ways. What are stocks worth as prices change seasonally and with inflation? Sometimes market prices are used,

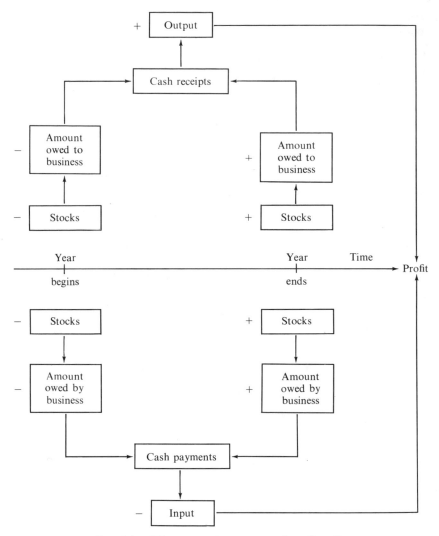

FIG. 6.1.   Diagrammatic representation of profit.

in other cases estimates of cost of production. Both can be wrong. For example, if a good has been in stock for some time its market price may be much lower or higher than the cost of purchasing it at the time of valuation. The use of current market prices may ignore any physical deterioration. Likewise, if a good is valued according to estimates of its cost of production (this will usually include arbitrary assumptions) there is no reason why this should be equal to the market price.

Two further unresolved accounting debates concern the treatment to be given to assets which appreciate and depreciate in value. Should, for example, the appreciation of breeding livestock due to inflation add to profits especially when in periods of rapid inflation the figures can be very large? Likewise, depreciating capital items on a historic cost of purchase basis can seriously understate the fall in value and therefore overstate profits. It also produces financing problems when and if replacement takes place. The alternative method of using replacement cost as the base for calculating depreciation assumes implicitly that the same item exists and will be replaced.

It is not always easy to separate out the various components of receipts and expenses into trading, personal and capital items. The farmer who travels to market may reasonably be expected to charge some of the costs of this travel to personal expenses, in so far as he spends some of his time there in leisure pursuits or on domestic matters. Again, expenditure on building repairs may, if it is substantial, be looked upon as the replacement of capital while for various purposes it may be treated as a trading item. Even commodities like fertiliser which are normally treated as trading items do have a significant residual value which could be allocated to capital.

Profit can be calculated in different ways. It is in economic terms the reward to management for risk-taking. It may be calculated in such a way that it also includes the reward to the farmer and his family for unpaid labour and for his investment of capital depending on what is included or excluded.

Wealth

The wealth of a business is measured by a separate and different device known as the balance sheet. The trading profit is a flow concept. The balance sheet is a static one and measures the capital structure through which trading activities are made possible. It is normally calculated at the end of a financial year but it can be constructed at any time. It is a listing of the assets and liabilities of a business (see Fig. 6.2).

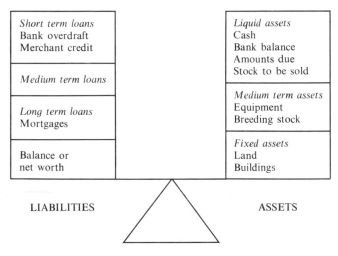

FIG. 6.2.   The form of a balance sheet.

The assets consist of stores of cash, unsold goods, unused inputs, equipment, breeding livestock, buildings and land. The liabilities are the calls on the business or loans made to the business by banks, by mortgage companies, by individuals and by the owners of the business. By definition a balance sheet must balance and the balancing item which equates the assets and liabilities should it occur on the liabilities side shows that the business is solvent. In other words, if all the liabilities were paid off through the sale of the assets some positive sum would remain. This sum represents the wealth of the owner of the firm, be he an individual or a set of shareholders. Insolvency means that the balancing item is on the assets side of the balance sheet. That is, the liabilities cannot be met in full by the sale or liquidation of the assets and in most countries of the world such a situation would mean that legally the business should cease to trade.

Everything that happens to a business can be expressed by changes in the balance sheet.[3] If something is sold out of stocks then the composition of the different types of assets will change. More cash or money owed to the business will take the place of an equivalent fall in stock values. If output is produced this will increase the amount of unsold products and reduce the stocks of inputs. These inputs may have been paid for in which case cash reserves will fall or alternatively the level of short-term loans will increase. Asset revaluation and profits which are not spent will add to wealth.

Profits generate business growth. They are reflected over time in the balance sheet either in the form of enhanced reserves of cash, larger stocks of goods or greater amounts of investment. Expansion which is financed entirely out of profits will be matched by an increase in the balancing item, sometimes called the net worth or equity of a firm. Profits can be consumed so that the wealth of the business may not increase at all. Indeed wealth may diminish if consumption exceeds profits and begins to erode the assets of a business. It is therefore important to examine the statements of profit and wealth together in order to evaluate the interactive interdependence of a business.

Profits contribute to the capital structure of a business by financing investment directly or by servicing loans. The capital structure of a business in turn is a major determinant of future profits. There is a dynamic interaction in the progress of a business between trading activities, consumption and capital investment.

A balance sheet can be used to identify the dangers of a serious financial problem known as overtrading. If overtrading takes place too great a proportion of assets are in forms such as buildings, equipment and machines with insufficient funds being available to buy inputs of a more variable character. The situation arises because management in attempting to grow more quickly invests funds required for working capital in fixed assets. While this is a serious situation in itself, it can lead on to an even more serious one in that customers of a firm begin to lose faith through delays say in the payment of bills. If confidence is lost then credit terms will become either more expensive or may be withheld altogether which only adds to the overtrading problem. Thus a solvent business can be forced out of business because the relationship between trading assets and fixed assets gets out of balance.

Errors in financial management normally first appear as negative cash flows. That is, receipts do not cover expenses over a period of time. In consequence, businessmen keep a careful watch on their bank balance or in the case of organisations operating on a fixed budget, on their total expenditure in relation to their allocated expenditure. This process is facilitated by using a simple cash book and records of what is owed and what is to be paid. Judgements are also assisted by being able to compare the level of expenditure with the historic pattern, for example last year's level, or more realistically with budgeted income and expenditure. Budgets may be completed annually or may be done on a rolling basis; that is they are updated periodically as events occur. Nevertheless, concentration by management solely on cash flow can result in errors since a positive cash flow can be maintained for quite a considerable time by increased borrowing or by realising capital. Too much cash in a business can also result in a loss of earnings. A holistic view of business finance is required and to achieve this accounts must be considered simultaneously.

The construction of accounts for management purposes begins with the careful recording of financial transactions in a cash book. Information is obtained from these data by aggregation and manipulation according to accepted rules and definitions. The usefulness of recording is clearly demonstrated by the insights accounts can give to managers.

Comparison

The significance of events is often judged by comparison with other events. The usual comparison is that of change with time, as reflected in the production of statistics which measure rates. Changes in the environment are measured by rates of inflation, indices of production and costs, wages and prices, output and productivity. Internal measures include changes in rates of growth, in sales and turnover. One of the tricks of management is to remove the inaccuracies of measurement by focusing on changes.

Provided consistent errors are made rates of change are accurate enough indicators of trends. Randomness and irregularities can also be smoothed out by using moving averages and adequate sample sizes.

The use of ratios in comparison removes the effects of differences in size. Financial ratios such as the various groups of costs expressed as a proportion of output tend to be stable for similar types of firms and over time for the same firm. Presumably this stability reflects the way in which compensatory mechanisms work within systems. Ratios are highly abstract measures of performance and they are therefore a most useful way of getting a comprehensive grasp of a business situation without the variety of more detailed information. They help to avoid the danger of jumping to conclusions on the evidence of detailed partial measures of managerial performance.

Ratios in common use which are calculated from a profit and loss account include variable costs, fixed costs and finance charges all expressed as a proportion of output or sales. Differences between sales and cost categories are also quoted such as the gross margin which is roughly defined as sales less variable costs. Ratios are also calculated from a balance sheet. The ratio of current assets to liabilities indicates the liquidity of a business. The proportion of assets which are owned by a business measures stability, and current assets expressed as a proportion of other assets give an indication of the danger of overtrading.

Margins and ratios are most helpful to management in identifying problem areas. They are too general to provide anything other than clues about the causes of problems nor is it possible to fix an ideal level for most of them.

Comparison can be misleading if consistency is not observed. This is sometimes not easy to achieve as for example in the calculation of yields per animal and stocking rates. The number of cows to be used in estimating annual milk yield per cow depends on the way in which cows are defined and how changes in the number of animals over time are worked out. An average of the opening and closing cow numbers can be misleading if the herd is contracting or expanding. This problem is particularly acute where businesses express their performance in league tables. An easy way to achieve status is to present figures in the best possible light.

There are so many different factors affecting performance that the definition of a sample of firms for comparison invariably involves assumptions about homogeneity. Accounting data, for example, cannot give an indication of the effect of qualitative factors such as managerial capability, differences in attitudes to risk and favourable environmental conditions. The comparison between individual results and a group average may therefore not mean much at all. High performance is more likely to be achieved in favourable conditions and successful managers will invariably operate large businesses. This type of information cannot suggest profitable changes to the small average performer operating in poor conditions. It tells him what he is.

Index Numbers

A common method of comparing changes over time is through the use of index numbers. Time series of prices, production and input use are expressed as a proportion of either a single base year or a series of years. Prices are often deflated for inflation, that is they are expressed in real terms. Great care is needed in the interpretation of index numbers.[4] Firstly, the base year or years may have been abnormal. Secondly, inappropriate deflaters can be chosen as, for example, when the retail price index is used for a category of goods whose rate of inflation is consistently different. The concept of the constant purchasing power of money is open to misuse because the basket of goods which people buy with their incomes changes over time. Thirdly, other weighting errors can arise when abnormal results have a distorting effect on a series of data such as, for example, where a few items make a high price or the inclusion of a few large firms distorts estimates of average incomes.

Comparing Incomes

An oft-stated objective of agricultural policy is to equate incomes derived from agriculture with those of the rest of the community.

Again great care needs to be taken in comparing like with like, since incomes vary by region and according to the demand for products which people produce. Moreover the purchasing power of incomes reflects the costs of living which again are not uniform throughout an economy. Applying a simple criterion of equating the income of one group of people to an average income can be an exceedingly clumsy and wasteful tool.

RELATIONSHIPS

The study of relationships uses data to identify their existence and to describe them in quantitative terms. Records help in all stages of developing knowledge about relationships. They assist in the association of events which through observation are thought to have either a causal or an empirical connection. Keen observation is a necessary skill in any form of management. In production processes, timely observation of changes in the status of crops, animals and machines can have a great influence on performance. Problems in the management of staff can be anticipated and detected by close observation of slight changes in behaviour. The making of notes about everyday occurrences in a diary or a formal log is an invaluable aid to observation.

Agricultural Research
A feature of the agricultural industry is that firms are insufficiently large to invest sufficient capital in research on their own account except for some large ancillary firms such as chemical and feed companies. The importance of research facilities in the development of agriculture has been demonstrated many times, as for example in the discovery of the importance of lime and phosphate and the application of fertilisers to crops. New varieties and developments in animal breeding and feeding have stemmed mainly from agricultural research.

The dissemination of relevant research results from Government-sponsored research institutes is an important and difficult task. One difficulty is to create a two way flow between the industry and the research centres. Another difficulty stems from the fact that relationships in agriculture are of a multi-variate nature. The response to a new practice in a particular situation will depend upon the circumstances that prevail. Fertiliser response, for example, depends on the variety of crop, features of the soil such as its moisture-holding capacity, chemical status and structure, the timing of operations, the incidence of disease, the sequence and onset of critical weather events such as rain, sun and wind, together with the success of management in observing and controlling crop development. It is impossible to reproduce the wide variety of circumstances to be found on farms on a few experimental sites even though they are repeated on different soils and in different years. Research results from experimental stations have to be treated more as hypotheses by individual farmers and their advisers who need to verify relationships for themselves before basing decisions on them.

Farm scale trials and demonstration plots help to test these relationships in situations nearer to existing practice.[5] The emphasis of such demonstrations is to reassure farmers of practical issues rather than to replicate treatments. For example, are new varieties acceptable to consumers in terms of taste and do they fit in with current husbandry practice and the local climate? Fertiliser response for crops grown as pure stands may not be relevant where local labour supply and weather events mean that mixed cropping may be more economic. A farmer has to build into his existing system in a profitable way any new knowledge that an experimentalist provides him with. The experimentalist needs to be sure of the relationships he has discovered and confident that there are no detrimental side effects. Unfortunately the two sets of objectives are not completely complementary. The farmer is primarily interested in all the interacting influences on performance and while he needs to be sure of relationships he may not need to be as sure as the research institute whose reputation would suffer if it were accused, by even a few people, of producing bad information. Unfortunately for research workers, size and cost of experiments which replicate a large number of treatments on different sites and over several years can soon get out of hand.

Even so, the farmer's problem is that he is interested in applying levels of new inputs which pay rather than simply knowing that an input produces a significant response. Incorporating different treatment levels within experiments to provide this information simply exacerbates the scale and cost problem.

Surveys

Another way of learning about relationships in agricultural systems is by means of surveys of practices in a population of animals, crops, farms or institutions.[6] It has the great advantage that the records reflect actual performance within the totality of all the influences that prevail. A disadvantage is that differences between firms cannot be entirely accounted for because of unmeasurable quantities such as managerial ability. Unexplained residuals in multivariate equations can be large and given that there is often correlation between assumed independent variables the functions derived from the survey data can be unreliable. Data available from surveys are commonly extremely general which means that while variation in performance may be explained say by differences in stocking rate, feed and fertiliser levels, it is not possible for the analyst to say how differences in performance are achieved. Information of a detailed nature such as the timing and type of inputs, the method of application and prevailing environmental conditions are usually not recorded.

Agricultural practice tends to converge. It is rare, for example, to find a wide spread of input levels between firms. There is either no measurement or at best only a few measurements available for extreme values. This adds to the statistical problems in this kind of work. A great deal of development has been successfully done on statistical methodology for handling survey data but the main limitation is the lack of data and the cost of collecting it.

Surveys can be expensive. (A Nigerian cocoa farmer study cost £70000 in 1953.[7]) A random sample of a farming population if it is to be sufficiently large to give reliable information on a large number of variables will run into several hundred farms. The travelling costs alone will be high for such a large sample especially if the farmers live in remote and sparsely populated regions where roads and other forms of communication are undeveloped. Telephone surveys, postal censuses and question-naires are only feasible in well-developed societies. The other main cost of surveys is that if specific data are to be acquired in an accurate way then it will be necessary for survey workers to measure or at least supervise farmers in the measurement of yields, areas, input levels, time and so on. Without such measurements the investigator will have to rely on the farmer's accuracy of recall. Several visits may need to be made if measurements are to be undertaken. Enumerators may have to live on farms if records of operational features of farming are to be made.

Securing a manager's cooperation in obtaining detailed data which will not produce any perceived benefit to him is a difficult and sometimes dangerous task. Some types of information are regarded by the agricultural community as politically sensitive and questions about land ownership, taxation and income levels can meet with positive resistance. This means that it is often impossible to obtain data of this kind by means of random surveys. The use of such information is usually restricted on the grounds of confidentiality.

There are many established procedures for minimising the costs of surveys such as roadside surveys, stratified samples and samples of villages rather than holdings, which reduces transport costs.[8] Important variables such as yields can be estimated according to derived relationships between what can be easily measured and the final quantity.[9] New developments in remote sensing can potentially give extremely precise and up to date information about land use.[10]

The expense of surveys means that data are often used for purposes for which they were not intended and this is turn leads to schemes to set up data banks to answer foreseeable and unforeseen questions. Unfortunately such schemes can deny the principle incorporated within the scientific method of collecting data with a clear null-hypothesis in mind.[11] Thus, accounting data which are collected for taxation purposes must be modified and used with great care when analysed for management purposes. Similarly milk yields collected for dairy cow breeding purposes have within them sources of bias which make them unreliable guides for decisions such as feeding.

There are also some practical measurement problems which need to be recognised when conducting surveys. Local measurements of volume, area and weight are often not accurate. They have a large standard deviation which could be exploited according to the interests of the informer. A housewife taking her basinful of grain to be milled obviously would prefer the basin to hold as much as possible if the charge is made for each basinful. Managers who receive productivity bonuses have an incentive to see that the criteria by which they are paid work out to their advantage. Where yields per animal or unit of land, stocking rates or prolificacy rates are quoted it is necessary to check how the number of animals, for example, is arrived at. They vary with time. Output is not instantaneous and estimates of land areas that are available for cultivation and are actually cultivated and eventually harvested, are all different.

Work rates for men and machines are even more difficult to derive since the available time for work which is influenced by the weather and field conditions is never fully exploited because of breakdowns, travelling time, rest periods and delays. In family situations it is quite likely that no records of precise hours at work are kept, while the potential output for work will vary between individuals according to skill, training, experience, sex, age and state of health. It is therefore unfortunate that so many measures of system performance are accepted without question and are used in inflexible ways by bureaucracies in development schemes and projects. Indeed it is a fact of life that there is so much variation within types of population, be they farms of a given type, breeds of animal, types of labour, that it is most difficult to differentiate the effects of changes in management. Even for relatively uniform enterprises such as pig and poultry production there are such striking differences between farms operating similar systems of production that improvements in management seem not to be a matter of the choice of system but of operating in a more efficient way the system which is actually chosen.[12]

Surveys as in any form of investigation can only be useful if the right question is posed and it is asked in an appropriate way. A common request is to work out the unit cost of production of a single commodity. Unfortunately, the lack of a system's view inherent in such a question forces arbitrary decisions on the researcher since he is obliged to allocate fixed costs and ignore the interactions between commodities including the obvious fact in many situations that output is produced jointly. Even for relatively independent enterprises there will still be problems of allocating overhead and management charges. Working capital requirements, for example, reflect the pattern over time of the receipts and expenses from the separate components of a business so that charges for capital can only be accurately estimated within the context of a whole business.

Such procedures might best be avoided if the original question to be investigated by the survey were put in a different way. Thus, if the cost of producing a commodity is required for price-fixing negotiations then a better question would be 'how much do relative prices have to change to alter the level of output?'. A survey to answer this question would try to ascertain the resource requirements of producing an extra amount of output. It would attempt to identify the more serious limits to production, producers' expectations and investment intentions. Time series of production levels and prices might be collected and analysed in order to give an idea of historical rates of response. New available technology would be appraised in the context of farms as a whole and from a national and regional point of view.

One of the more serious omissions of data on agriculture is the predominant use of the arithmetic average. Yet a universal problem for managers in agriculture is how to cope with uncertainty in both the physical and economic environment. For those intent upon survival it is imperative that they anticipate the possibility of both good and bad events occurring and make estimates both of the extent and effects of such events together with their likelihood. Variation in technical relationships can be exploited by management and to ignore it will prove disastrous. Published estimates of such variance are rare and not too reliable since the time period over which technology is unchanged within a firm is often quite short. It is not logically possible to estimate variance on technical relationships using data from different firms. However, by utilising the available knowledge on causal relationships it should be possible to at least determine the bounds of possibility and where knowledge is well developed to even derive a frequency distribution using simulation methods.

Surveys in summary have three major benefits. Firstly, they are most useful for providing a description of the agricultural industry. Policy measures by governments and investments made by farmers and entrepreneurs can be put in context while discernible trends are helpful in the formulation of expectations. Those who doubt the value of surveys should contemplate the quality of decisions made in the absence of information provided by surveys.

Secondly, they can provide data for planning purposes but because surveys are expensive to conduct and are confined to general measurements, relationships are hard to discover especially when in cross-sectional studies there are large unexplained differences between firms. However, they do present challenges to management in that the possibilities of improvement are clearly evident by a comparison of relative performance.

Thirdly, specific surveys on practices, social behaviour, disease incidence, soil and plant nutrient status, market trends and general expectations can, when correlated with each other or with measurements of output or input, provide good evidence for the existence of relationships. This may, where knowledge is sufficient, give some guide on the quantitative impact of changes. In so far as this kind of knowledge is based on actual practice then the data are likely to be more realistic than those gleaned from purely experimental evidence. Both kinds of investigation preferably designed in association with each other to answer well-defined questions are necessary.

REFERENCES

1. *Annual review of agriculture*, 1982, Her Majesty's Stationery Office, London. Cmnd. 8491.

2. NORTHCOTE PARKINSON, C. and RUSTONJI, M. K. (1975). *All about balance sheets: the easy way*, Macmillan, London–Basingstoke.
3. BOULDING, K. (1950). *A reconstruction of economics*, Wiley, New York.
4. WALLACE, W. H. (1974). *Measuring price changes: a study of the price indexes*, (2nd Ed.), Federal Reserve Bank of Richmond, Virginia.
5. DE DATTA, S. K., GOMEZ, K. A., HERDT, R. W. and BARKER, R. (1978). *A handbook on the methodology for an integrated experiment–survey on rice yield constraints*, The International Rice Research Institute, Philippines.
6. KAHLON, A. S. (1975). *Impact of mechanisation on Punjab agriculture with special reference to tractorisation*, Department of Economics and Sociology, Punjab Agricultural University, India.
7. GALLETTI, R., BALDWIN, K. D. S. and DINA, I. O. (1956). *Nigerian cocoa farmers: an economic survey of Yoruba cocoa farming families*, Oxford University Press, London.
8. BESSEL, J. E., ROBERTS, R. A. J. and VANZETTI, N. (1968). *Agricultural labour productivity investigation. Survey field work*, Report No. 1. Universities of Nottingham and Zambia.
9. BECKETT, W. H. (1972). *Korangsang cocoa farm 1904–1970*, Technical Publication Series, No. 31. Institute of Statistical, Social and Economic Research. University of Ghana.
10. STOVE, G. C. and HULME, P. D. (1980). Peat resource mapping in Lewis using remote sensing techniques and automated cartograph, *Int. J. Remote Sensing*, 1 (4), 319–344.
11. JEFFERS, J. N. R. (1975). Constraints and limitations of data sources for systems models, In: *Study of agricultural systems*, Dalton, G. E. (Ed.). Applied Science Publishers Ltd, London.
12. RIDGEON, R. F. (1981). *Pig management scheme results for 1981*, Agricultural Enterprise Studies in England and Wales, Economic Report No. 80, Department of Land Economy, University of Cambridge.

CHAPTER 7

Forecasting

Actions in the present produce outcomes in the future. These outcomes depend on the initial state of the system, the way in which changes in the rest of the system influence performance and the effect of environmental changes. The outcomes occur over time so that forecasts depend on the time period being considered. The accuracy of forecasts depends on the correct perception of the major influences on an outcome, the response of the system to these influences and which events occur.

Forecasts are made and used in management for planning, control and day to day operation. A simple case is the use of a weather forecast for deciding on the jobs to be done, on and within a particular day. Forecasts of, say, the rate at which men and machines accomplish certain tasks are used to plan and monitor job progress. Production targets can be based on forecasts so as to monitor and control the performance of animals such as milking cows, pigs and poultry. Forecasts of prices, production, available supplies and consumption are used in decisions which determine, among other things, what and how much is produced in the future. Thus, farmers use this kind of information to decide on the area of particular types of crops to grow. Governments use it to decide on subsidy levels, quota limits and import controls. Longer term forecasts are required for

TABLE 7.1

Outlook Work—Purposes, Information Requirements and Dissemination Methods

Time scale	Typical management tasks	Information requirement	Method of dissemination
Short term	Tactical	Highly quantified and specific	Personal contact
	Buying-selling of grain; storage; cattle buying and marketing, feed and fertiliser purchase.	Forecasts of daily, weekly and monthly prices and price trends.	Telephone
			Media—viewdata
		Futures prices.	Information leaflets
	Buying-selling futures.	Physical and financial data on farm.	Newsletters
	Managing overdraft.		
	Technical management	Timing of fertiliser application and appropriate rates; spraying disease control	Discussion groups
Medium term	Tactical/strategic	Medium to high level of quantification	Outlook conferences and proceedings
			Outlook bulletins
	Breeding and stocking	Seasonal movements in prices and changes in overall price levels	
	Culling and replacement	Animal populations, cropping areas	
	Cropping—grass proportion, undersowing, etc.	CAP determinations	
	Enterprise mix changes (particularly in autumn)	Forecasts of prices/gross margins	Economic management information bulletins
		Budgets for representative farms	
	Production quality, e.g. seed/ware, malt/feed.	Forecasts of price premia for quality differences	
	Fattening-feeding strategy		
	Equipment replacement and renewal	Interest rates, machinery prices, returns on output	Technical bulletins

Technological innovation Legislation—taxation Direct outlet to retail, packaging and presentation, etc. Export orientation (EEC)	Recent developments on market: round bales, growth promoters, Tractor cabs, investment incentives VAT—changes in rates, etc. Strength of market demand—retail prices; market situation in Europe (e.g. lamb)	Discussion groups Farm case studies
Accounting systems/office organisation	Allowance for inflation in asset valuation	Lectures and articles to agricultural societies and groups
Long term Strategy policy Large scale capital investment (buildings, land, etc.)	Degree of quantification low Indicative forecasts Changes in world economic growth, trade	Outlook conference and proceedings Occasional studies
Intensification of existing systems Enterprise balance Scale of enterprise Specialisation	Supply-demand balances in developing and developed countries for commodities and inputs—movements in relative prices and costs	Farm management reviews
Changes in marketing policy	Forecasts of structure of wholesale-retail markets	Lectures and articles for general as well as agricultural societies and groups

major investment decisions such as irrigation schemes, building projects, land improvements and the installation of processing facilities. Table 7.1 shows the usefulness of forecasts of different types to farm managers and ways in which this information can be disseminated.

One way of making forecasts is to extrapolate from past behaviour on the assumption that trends, cycles and seasonal patterns will persist. A special case of this method is the assumption that present conditions will continue. In fact, it is quite likely that more immediate occurrences have a greater influence in the formulation of expectations about future states. A different type of forecasting method is to study the correlation between different historic events. Such studies can result in the establishment of laws or relationships which assist in the interpretation of future events once they occur even if the events themselves cannot be predicted.

THE NATURE OF FORECASTS

Consider a situation where a farmer is contemplating buying some lambs for fattening. The eventual profit or outcome from his decision will partly depend on how many he can afford to buy. The number of animals he buys will depend on how much money he has, the price of the animals and the quantity of feed available. These are all initial conditions. Another less obvious but important starting point is the number and kind of alternative opportunities available to the farmer. The farmer in order to decide on whether or not to buy the lambs will need to make a prediction of the outcome which will be defined in this example as the revenue received for the animals when they are sold less the costs of buying and keeping them. The revenue will partly depend on how well the animals grow. The feed costs will be a function of the amount of supplementary feed the lambs eat. These production forecasts of growth rates and feed consumption can be calculated from the extensive knowledge that exists about animal nutrition. However, three sources of uncertainty can be identified in making such forecasts. Firstly, the farmer's knowledge about nutrition is likely to be incomplete. He may not even know about it or if he does he may use it incorrectly. Secondly, the nutritional information will be subject to error. Thirdly, the data required by the farmer to predict lamb growth, such as the initial weight of the animals, the quantity and quality of the feed available, may be only crudely estimated. Further, feed requirements may depend on the weather; unfavourable weather reduces the amounts of feed available and increases the amount of food required for a unit increase in weight.

Given that the animals survive, grow and become fat, it is necessary to estimate the final price in order to calculate the returns. The level of prices will normally be outside the farmer's control. He may achieve quality standards which give him a slight premium over the average price and he may be fortunate enough to receive some form of Government guarantee. In the absence of these measures he is at the mercy of the market but as with weather events the farmer is not completely ignorant about possible future states of the market. Indeed he may feel quite confident about the level of future prices. His confidence may be based on his past experience or a rule of thumb which he believes has served him well on previous occasions.

This example shows that forecasts are made up of several fundamental parts. Some predictions can be made with greater confidence than others but even these are subject to errors caused by inaccurate data and ignorance of relationships. The eventual outcome is also conditional on which environmental events occur and how they influence the state of the system. Forecasts, therefore, can rarely be single point estimates. A wide range of future conditions are possible. Forecasts are used to make decisions and the beliefs held by the decision maker rather than the professional forecaster are what matters.

DECISION TREES

A systematic way of representing all the separate assumptions in predicting outcomes is known as a decision tree. It consists of a network diagram where acts and events are drawn in chronological order. Only acts and events which the decision maker considers to be relevant are included in the diagram. The representation is therefore only as accurate as the decision maker's knowledge. A decision tree for the example described in the last section is shown in Fig. 7.1.

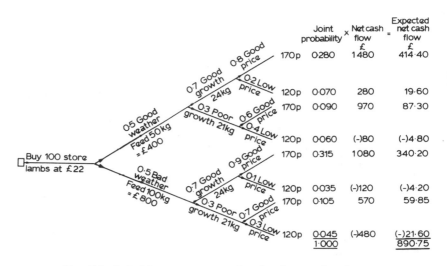

FIG. 7.1. A decision tree representing the elements in a forecast.

The array of possible outcomes in the lamb fattening example ranging from a loss of £480 to a profit of £1480 is a feature of uncertainty. Managers in order to make decisions have to be able to compare such arrays with those from other activities. A necessary part of such comparisons is to reduce the variety of the forecasts to a more manageable number by, for example, weighting each individual outcome by the manager's feelings of the likelihood of particular events occurring.

Statistically, the quantification of likelihood is invalid unless it can be safely assumed that there is a large universe of like events and that the conditions which affect the chances of an outcome occurring remain the same. Some types of events approach these conditions, for example, the risk of fire or the performance of a particular genotype. What is not known even for this type of event is the performance of a particular individual within the population. While an insurance company can specify the odds of a building catching fire, which building will actually catch fire is not known. The likelihood of rare events such as booms and slumps, technical developments and political initiatives cannot by their very nature be described by frequency distributions. Thus there is no satisfactory objective way of measuring the degree of confidence managers feel about unusual but important events.

The derivation of production relationships in experimental programmes is carefully done so as to allow the statistical testing of differences between treatments and relationships between variables. Estimates of the chances of obtaining responses can be made by such means. However, the published results are usually only those which conform to fairly high levels of significance irrespective of the potential value of the information to the manager. Further while a response may be estimated to occur on most occasions, the significance level only applies to the data generated by that particular experiment. There is always an added risk that the conclusions from an experiment may not be repeatable in a real situation. This risk may be minimised by repeating experiments in several different situations.[1]

The feelings of confidence that managers have about the future are sometimes handled according to the rules of probability. The term used for expressing likelihood in this way is that of subjective probability. The approach assumes that decision makers can measure differences in their degree of confidence and that they can state their quantity of confidence. Procedures have been developed to estimate subjective probabilities using introspective questions about imaginary bets. Such precision can be misleading. Many people find it difficult or impossible to differentiate the impact of an event from its likelihood of occurrence. We all indulge in wishful thinking so that the intensity of feeling about the future will be made up of both the size of the outcome as well as the likelihood of its occurrence as behaviour at race meetings illustrates. The excitement of a race meeting is an amalgam of the fun of guessing which horse will win, augmented by the size of the bet and the quoted odds.

Two approaches for simplifying forecasts can be demonstrated using the decision tree shown in Fig. 7.1. If it were possible to assign subjective probabilities to each event then each outcome could be weighted by the joint probability of its occurrence. The sum of the expected values can then be calculated so that the array of outcomes is reduced to a single number. This can greatly assist in the comparison of alternative actions.

Another way which has been suggested for reducing an array of outcomes to a more manageable number is through the use of certainty equivalents. In Fig. 7.2 if it were possible for the decision maker to state a quantity for certain which he would be just prepared to accept in exchange for an 80 % chance of £1480 and a 20% chance of £280 then the complexity of the decision tree can be reduced. Proceeding with this process would result in a unique certainty equivalent for the whole tree.

A more realistic approach for simplifying forecasts can also be demonstrated from the decision tree in Fig. 7.1. The farmer may choose to concentrate on those events which he is confident about and ignore the rest. Thus the joint subjective probability of good weather, good growth and a low price is only 0·07 and he would choose to ignore this on the grounds that it is so unlikely as to not be worth consideration. Proceeding with this kind of analysis only three events may be contemplated, namely, good weather, good growth and a good price; bad weather, good growth and a good price; and bad weather, poor growth and a good price. In other words the farmer may on the basis of past experience feel sure that he can buy animals and manage them in such a way that they will grow well. While he may feel that low prices are a possibility he thinks the chances of them occurring are sufficiently slim for them to be ignored and in any case even if low prices occurred provided he gets good growth he would not make an absolute loss. Because he cannot predict the weather then he ascribes an equal possibility to the chances of good and bad weather occurring. The extra feed required in the bad weather will not ruin him. Thus, by ignoring

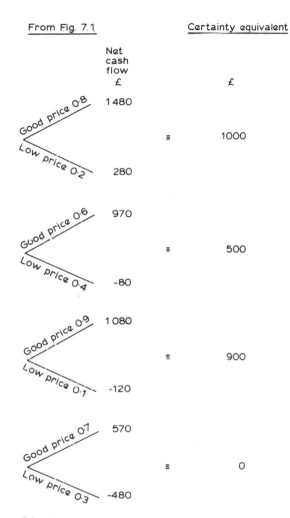

Fɪɢ. 7.2. The use of certainty equivalents to simplify forecasts.

the unlikely and the unimportant in his forecasts the farmer has significantly reduced the complexity of his decision.

Another classical response to uncertainty is to wait until the future is nearer and clearer. The results from elections become more predictable as they draw near. Price negotiations are concluded and experience is built up as new technology is tried out. By waiting, more information is acquired and confidence is increased. It is possible to view the role of agricultural research and extension activity in this light. Additional information generated by such services modifies or augments the initial beliefs that decision makers in the industry already have. This type of role is often denied on the grounds that assessments of likelihood other than objective ones are irrelevant. It is important to remember that decision makers do not behave according to one set of information. In the end it is how they feel about likely outcomes which determines their actions. For example, it has been found in recent studies that farmers apply pesticides to crops according to their feelings about the likely response and in some cases these feelings were inconsistent with known facts.[2] The degree of uncertainty in agriculture for many decisions is such that there is no alternative but to base decisions on anything other than hunches.

LIMITS ON PROSPECTS

While humans have freedom and the physical environment is not understood completely and is not controllable, then unexpected events will always occur. The range of outcomes is, however, restricted by physical limits, existing conditions and capabilities. A particular concern in agriculture is the growth in yields. Linear or exponential curves can be fitted to historical data and extrapolated through time. The type of curve fitted should reflect the knowledge that exists about biological limits to physical yields. Linear or exponential extrapolation soon becomes unrealistic.

Chartists and decision makers preoccupied with trends, sequences, seasonal and irregular movements in prices, production and costs, can easily miss the obvious. In the early 1970s for example livestock populations were built up in most Western nations on the back of an economic boom which collapsed in 1974. With hindsight it is possible to discern that even had prosperity continued there was not unlimited demand for meat consumption and that high expectations were feeding on themselves through supplies being restricted in order to build up the livestock population even further. What goes up must come down is all too true but it is more helpful to understand the reasons. It is necessary then in forecasting to concentrate on the important parameters which circumscribe the ranges of reasonable probability. National agricultural plans can overlook the fact that insufficient land is available to meet increased production targets unless unprecedented levels of performance occur. In the lamb fattening example shown in Fig. 7.1 it is unlikely that the lambs will exceed the mature weight for their breed.

Calculations to check these are not strictly forecasts, rather feasibility studies which show that existing trends cannot continue without modification.[3] Discontinuities exist within systems. They include limits on size, the availability of resources and rates of response as in growth and other biological and physical transformation processes.

The future performance of any system is also limited to a large extent by its existing state. A thorough situation analysis can be most informative as preparation for a prediction exercise. The area of a particular crop or the numbers of livestock largely predetermine production for the coming season. They also constrain the change in the succeeding year's production because of rotational constraints, biological limits on population increases and the availability of knowledge and the capacity of facilities. The age, quality, number and health status of tree crops, for example, constrain future production levels for many years while for livestock with long breeding cycles the same is true. The financial state of the agricultural industry also determines what can be done in the immediate future. Unfortunately the information that is available on liquidity and profitability is often out of date so that bad decisions are made because changes that have occurred are not recognised. Investments continue to be made when booms have ended and when profitability is less than it is thought to be.

Understanding how a system works increases the confidence

attached to forecasts. For example, the profitability of investment decisions in a free market system depends in general on achieving above average levels of performance. If this were not the case then the supply of commodities produced by the investment would need to be in shortage. If shortages are not present the profitability of adding to supplies is a matter of beating the competition through lower costs. A budget which shows that an investment is worthwhile can therefore be checked by comparing anticipated efficiency with that of existing and potential suppliers.

An even simpler case of the value of understanding a system is that of visually estimating the weight of an animal. This can be done in several ways. Those with experience can compare the animal in front of them with animals of a similar size which have been weighed. They weigh by comparison. Studies have shown that the weight of an animal is closely associated with its height so that it would be possible to gauge the height of an animal and multiply up by the appropriate coefficient. Behind this relationship, however, there is the explanatory logic that the weight of an animal can be thought of as its volume times density. The animal is roughly cylindrical in shape and the volume of a cylinder is given by $\pi r^2 l$ where r is the radius and l the length. This shows that the most important determinant of the volume is the radius. The greater the radius of the animal the taller and the heavier it will be.

Models of systems produce a greater understanding of the way systems work. The combination of all the available knowledge about a system in a systematic way makes it possible by sensitivity analysis to discover which are the critical elements affecting performance. Models concerned with the growth of firms for example show the importance of the cumulative effect of high performance levels and acquiring capital according to how well the repayment schedule matches the availability of cash rather than the rate of interest. Models of harvesting systems show that the dominant cost influencing operational and capacity decisions is that of the losses to the crop caused by delay. Empirical studies of agricultural markets show that it is usual for supplies to be the main determinant of prices, at least in the short term.[4]

MANOEUVRING FOR POSITION

The decision tree also serves to illustrate the need for management to be aware of the danger of getting trapped in fixed positions. In Fig. 7.3, a situation is shown where choosing an initial course of action rules out the possibility of a subsequent choice. The fear of ruinous price wars, for example in markets dominated by a few large firms, explains why competition is normally based on the provision of services. Similarly, a small firm in a market dominated by much bigger ones may not expand since to do so would invite the attention of its competitors who may acquire or destroy it. There are routes in the decision tree which are not taken because the perceived action will invite an unwelcome event.

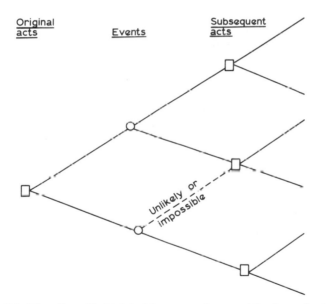

FIG. 7.3. The effect of initial decisions on subsequent freedom of action.

Similar logic also applies to a firm or a nation which chooses to indulge in a high cost system of agriculture. Such a system is more vulnerable to the natural fluctuations in market prices since profit is the difference between two large numbers and a small change in either will have a more than proportionate effect on the residual. A characteristic of decisions which influence the fixed costs of businesses is their irreversibility. If a large investment is made this will rule out other options in the immediate future. Since investment provides the framework for future production possibilities a wrong decision can produce serious consequences for a long time. Unfortunately, strategic decisions of this type can be made in the euphoria attached to windfall gains especially when coupled with tax incentives. Guarding against the installation of systems which are not sustainable in more normal times is an important function of forecasting.

REFERENCES

1. BOYD, D. A. (1979). Interpreting published information. *Farm Management*, **4**(1), 33–40.
2. MUMFORD, J. D. (1981). Pest control decision making: sugar beet in England, *J. Ag. Econ.*, **32**(1), 31–41.
3. MEADOWS, D. H., MEADOWS, D. L., RANDERS, J. and BEHRENS, W. W. (1972). *The limits to growth*, Universe Books, New York.
4. WRAGG, S. (1981). *Making decisions with the aid of market information*, UK Meat and Livestock Commission, Economic Information Service, Milton Keynes.

CHAPTER 8

Implementation

Attention to detail and the quality of work done are crucial to success in agriculture. Many different factors all combine to influence results so that the cumulative effect of even slight improvements in the execution of each decision can be large.

Day to day management in agriculture is circumscribed by the seasonal pattern of work and by the institutions that exist and their capabilities. Nevertheless, within this framework, the way in which people within agriculture are treated, priorities are set, time is used, and organisations are run can make all the difference. One major reason for this is the irreversibility of decisions within agriculture. Productive capacity is influenced to a large degree by the quality of assets that are available. Buildings can be badly designed and sited, fields can be small, farms fragmented, the soil can be eroded, people can be exploited, training and research irrelevant, markets poorly developed and incentives absent. Improvement can be difficult, expensive and slow once deterioration has been allowed to occur. Indeed experience with land reform and consolidation programmes shows that improvement can be almost impossible. In other cases if price incentives are sufficient, production can recover quickly despite organisational problems such as land tenure arrangements but physical assets necessary for production have to be made adequate.[1]

THE ANALYSIS OF PROBLEMS

Changes in day to day management are made in response to problems; that is, when performance is judged to be unsatisfactory. Unfortunately problems are frequently badly analysed so that wrong solutions are proposed and implemented. Solutions that are attempted will depend on the accuracy with which the causes and effects of problems are perceived. Thus in the treatment of a sick animal, recovery depends not only on the effectiveness of the cure but on the accuracy of the diagnosis. Treatment can only begin if the animal has first of all been identified as ill. This is not easy and depends on detailed, frequent and knowledgeable surveillance. The measures taken, their cost and effectiveness may well depend on the stage at which the illness is first observed. In all types of management, early detection of problems is usually helpful in that the costs of secondary and cumulative effects can be avoided. If the cause of the problem rather than the symptoms can be identified then future problems may be prevented from occurring.

An example from machinery management shows how cause and effect vary according to the point of view. If a machine breaks down, the task of the mechanic is greater than the repair job itself. First of all he needs to find out what is preventing the machine performing. He can only repair the machine if he knows this. His other task is to find out if the fault was caused by bad maintenance, bad operation, poor design or a manufacturing fault. A whole chain of causality can be built up. The effect of poor design is a faulty part which in turn is a cause of breakdowns, which in turn can cause high repair bills and so on.

A series of questions can be posed to assist in the definition of problems and to find out their causes and effects (Table 8.1).

The first question is to ask what has changed. Has a machine that has broken down been recently serviced? Has the operator or job changed? Has a sick animal had a change of diet; has its history differed in any way from the rest of the herd or flock? More precise specification can be achieved by seeking out the time and locational aspects of a problem. When and where did it occur

TABLE 8.1
Questions to Clarify the Definition of and Reasons for Problems

1. What is the problem as distinct from the cause and the effect of the problem?
2. What has changed?
3. What is affected—where and when does it occur?
4. What is not affected—where and when does it not occur?
5. What is the extent of the problem—how big is it? How often does it occur? Are the size and frequency increasing?

and just as important, when and where did it not occur? Are animals sick in a particular field or building or at a particular time of year? Do machines break down on specific jobs in particular areas or when operated by a specific person? The extent or size of the problem also needs to be determined in order to judge whether the problem requires attention immediately and how many resources are justified in trying to correct it. How many animals are sick? Are they seriously ill? How frequently do machines break down? Are they costly or difficult to repair?

Investigating a problem in this way helps to ensure that investigation is systematic and prevents management from jumping to conclusions. This is most difficult to avoid where several people or departments are involved. People can take up defensive positions and deny any knowledge of events associated with a problem especially if they believe a scapegoat is required. Investigation can easily lead to allegations that it is someone else's fault, be it administration, the marketing department or the customer. Differing and limited perspectives by separate departments make it difficult to avoid wrong conclusions being reached and to piece together an accurate description of a sequence of events associated with the problem. Several solutions are normally suggested without being sure of reasons. A check on conclusions is therefore helpful by trying to anticipate the effects if the proposed solution were in fact adopted. The cure can be worse than the disease. Past experience is invaluable at this stage. If the problem has occurred before, the action taken on previous occasions and its effects can be useful evidence provided the present situation is similar in all important respects.

PEOPLE

Job Satisfaction

Agriculture is a human activity. It is done by people in order to satisfy various requirements of society at large. Management invariably means getting work done by other people. Even the smallest farmer enlists the help of his family. How does a manager create the conditions by which other people will perform well? The motivation of people is not easy to accomplish or to write about. Generally it is true, if trite, to say that people should be respected as individuals, goodwill should be regarded as precious and freedom of expression defended. People are not machines. They have feelings and the purpose of agriculture is to serve people and not the other way round.[2]

Earnings and work conditions are naturally important in ensuring a motivated work force but several studies have shown that these are not the most important determinants. The ability to be creative, to feel useful and to achieve recognition through what is done tend to be ranked more highly than monetary rewards. There are also common reasons for worker dissatisfaction varying according to the organisation from the management, the supervisor, the Government or the administration. However, dissatisfaction does not necessarily mean poor performance.[3]

The interaction between individuals is also important in boss–subordinate relationships and in team work. The four general states derived from transactional analysis[4] are shown in Table 8.2. The ideal is that of feeling good about oneself and about the person being dealt with. The worst situation of feeling bad about oneself and critical of others can result in serious problems. The analysis does bring out the need in providing opportunities for people to work on and improve their strengths since this is likely

TABLE 8.2
Types of Attitudes between Individuals

	I feel	
About me	OK	Not OK
About you	OK	Not OK

to be much more rewarding than trying to correct their weaknesses.

The Purpose of Development

The need for people within agriculture to feel and be creative, to belong, or to have an identity, to be accountable and yet free, questions the whole purpose of agricultural development be it at a national, regional or institutional level. At the national level food of the right kind must be produced in sufficient quantity. Firms must produce profit to survive and public institutions must operate at an acceptable standard within their budgets. But measuring progress only in terms of these criteria ignores the fundamental reasons for development which are that the capabilities and choices of people are enhanced. Such a view expands the objectives of national agricultural plans.

One of the errors of national or regional agricultural development schemes which are supported by central or international finance is to impose the values of the donor by the way in which incentives are offered. Assistance, for example, which is conditional on income per head being below some so-called poverty threshold ignores the arbitrary nature of such measures and the fact that there is more to life than income. The importance of culture, freedom and social development is unlikely to be the same for both donors and recipients. One of the tempting pressures for all development agencies is to create the outcome they desire by their own efforts. Thus, for example, irrigation schemes are built and then farmed by Government agencies. Subsidies are heaped on subsidies as failure becomes apparent with the consequential growth in dependence, dishonesty and discord. Extension schemes can easily end up with advisers assuming managerial responsibilities, especially where technology is beyond the capabilities of farmers or inappropriate in some other way.

EQUITY AND DEVELOPMENT

The need for clarity and simplicity in bureaucratic procedures by urban policy makers can result in programmes that are intended to produce food but which bypass most rural people. An extreme example is the setting up of state farms in an area of small farmers which, while modern and mechanised, have very little direct effect on the local population. State farms deprive small farmers of land and provide wage employment in return. They tend to waste capital and foreign exchange and drive down prices for local products and thus deter development of this sector. The same sort of effect is also produced by blanket price support measures based on average costs of production. The above average performers grow and simply because of scale, which may also improve their efficiency, they receive a disproportionate share of subsidies—whose prime object is often to maintain the incomes of smaller farmers. Mechanisation is closely associated with this process. The large and successful can acquire machines which enhance their power to control more land. The result can easily be an agriculture of a few large farmers and many landless labourers with either uncertain seasonal employment or no better alternative than to migrate to urban slums.[5,6]

The difficult task of management is thus to create conditions where the undoubted physical benefits of mechanisation and other forms of technical progress are acquired, in the form of lower priced and adequate supplies of food, but where the social costs are not unwittingly heaped upon those least able to bear them. This can be achieved through land ceilings and taxation, including progressive income and wealth taxes as well as more discriminating means of support.[7] However, such measures are difficult to implement especially as a few large farmers have a great deal to lose and yet because of their influence are likely to be both political representatives of agriculture and at the same time are wooed by governments anxious to maintain food supplies for their urban power base.

EQUITY AND EXTENSION WORK

The conflict between cheaper food and equity also manifests itself in agricultural research and extension programmes. Agricultural research tends to be a profitable way for governments to invest their funds, the main benefits being bestowed on consumers through lower prices.[8-10] In so far as lower prices are of more significance to lower paid urban people and indeed to low paid rural communities the socially weighted benefits are enhanced. The dilemma for research and extension organisations is that the financial returns from gearing their programmes to the few large farms, who produce most of the output, are generally higher than a more industry wide approach, because costs are lower and the response to advice that is adopted will apparently be greater.[11] It is likely that there will be fewer problems of communication, both in language and attitude, and because fewer farmers are involved there will be fewer problems to tackle and understand. However, some of these problems can be avoided if smaller farmers are influenced by the diffusion process within agriculture, where new and profitable technology is adopted first by innovators who in turn are copied in their efforts through the incentive of innovators' profits and eventually, as more and more farmers adopt it, through competitive pressure.

COUNTERING UNFAIR COMPETITION

Exploitation of farmers in general is made possible because of their weak bargaining position in markets, especially in situations of oversupply. They are normally faced with a few buyers for their produce, especially in exporting situations. They inevitably end up being given a price which they cannot refuse in the absence of other markets. The response to this situation is for farmers to attempt to balance out the bargaining power by some sort of collective action.[12] Usually, however, some legislative support is required since the more successful a group is the greater the incentive for an individual to opt out. In so far as an individual's influence on the market is negligible, by opting out he can retain the rewards from the group action without incurring the costs. Such costs might include restrictions on output, quality standards and payment terms. Governments then often support cooperatives—they may be given monopoly buying powers or financial support through favourable loans or tax incentives. Indeed, there is a danger that such powers can be so effective that consumers can be exploited by farmers. However, where Government support is not forthcoming such collective action is generally not successful.

Another way of supporting farmers' incomes is to control supplies. This can be done directly through the imposition of quotas, or by organising the market so that one organisation controls all the supplies. This organisation, in a situation of oversupply, can discard excess production in such a way as not to disturb the market. One of the longer term effects of quotas is that progress in the industry is ossified so that consumers not only suffer the costs of restricted supplies but the longer run tendency for costs to be higher than they otherwise might be. One partial solution to this problem is to allow quotas to be saleable so that the more successful at least have a chance of either entering or growing in the industry.

The Ownership and Operation of Land

Who controls land is an especially difficult problem. If ownership and operation are separated, then without security of tenure the tenant has little incentive to improve the long-term performance of the land. Where share cropping is practised, many economists have argued that short term or variable inputs will not be used up to the point where the returns generated by the last unit will equal their cost since the returns are reduced according to the proportion going to the landlord.[13] This disadvantage may be offset by the provision of capital and the sharing of risk. In other words the situation is one of partnership.[14] Too great a security of tenure can make land ownership by itself unattractive which increases the capital requirements of the operator substantially and may restrict his progress in the industry.

The selection of individuals as farmers can be more a matter of luck than good performance. Inheritance laws may allow a complete transfer of resources to the next generation without regard to ability. If inheritance is taxed then this makes competition more equal. However, inheritance is an extremely sensitive topic. Farmers become attached to their land. For many it assures a food supply. In other cases it acts as collateral for loans or mortgages and is a symbol if not an actual source of status and power. The objective of many farmers is to pass it on to their families. A free market in land is thus hardly possible so that the selection pressure on who is to be a farmer can rarely be consistent. In the case of inheritance where all children receive a fair share then the consequences of this are a continual fragmentation of holdings if the population is increasing.

EDUCATION AND TRAINING

Improving the inherent skills of the agricultural population is a necessary part of progress. The capacity of people is often the most limiting factor to progress. Judged from an economic point of view, the returns from education and training should exceed its costs, including the largest cost item, that of time devoted to learning by the student.

One of the difficulties in this area seems to be the necessity for admistrators to distinguish between education and training. The question is superficial but in terms of what is taught, education is associated with more fundamental knowledge or the 'why' aspects, whereas training is concerned with knowledge of the 'how to do' kind applied to particular tasks. In both types of activity failure is guaranteed if the student is not required to think. Education that is irrelevant will not attract students and training that does not recognise the whole man is empty. Instruction in agriculture, whether of a general kind or specific, has to both educate and train and more importantly foster a spirit in staff and students alike of jointly searching for knowledge.

A difficulty of instruction in agriculture is the wide range of relevant subjects: science in all its forms but also economics, sociology, engineering, statistics, logic and so on. It is easy for comprehensive courses to become long and descriptive so that higher levels of learning are not achieved. Knowledge is not applied to problems and judgement is not developed. It is not necessary to cover all subjects to achieve these latter objectives or to go deeply into a specific subject area to acquire these skills. The ability to sort out relevant information from different areas of knowledge at an appropriate depth according to a correctly posed problem must be the aim of all instruction.

Task Analysis

Curriculum development is a continuous process, because needs and knowledge are always changing. A systematic approach especially in the design of training programmes is to analyse in so far as this is possible the tasks that trainees do or will do.[15] An illustrative example of this process for an adviser follows.

A task analysis begins by first defining broad areas of responsibilities or duties and then for each duty defining jobs and in turn elements of these jobs. An illustrative list of duties for an adviser might be:

(a) An adviser provides information on request to farmers and colleagues.
(b) An adviser solves technical and financial problems on request from farmers.
(c) An adviser produces plans for enterprise operation on request.
(d) An adviser produces plans for the whole farm on request.
(e) An adviser seeks (finds out about) new developments appropriate to the farms in his area.
(f) An adviser actively tries to discover appropriate new developments.
(g) An adviser evaluates appropriate developments.
(h) An adviser seeks to communicate knowledge to farmers and colleagues.

For each duty jobs can be defined. For example the duty of providing information means that the adviser must be able to:

(a) Keep up to date with technology
(b) Be able to find information
(c) Be aware of sources of information
(d) Be able to express himself clearly
(e) Be able to judge the relevance of the request
(f) Be able to judge when it is necessary to call on the services of a specialist
(g) Be able to choose the right specialist
(h) Be able to understand a client's request.

Each of these jobs will involve certain elements, e.g. (a) will involve reading the literature, observation, conference attendance; (b) and (e) will involve using the library, information retrieval systems, filing.

The next part of the training process is to assess the trainee against the list of the requirements. It may be that the most sensible policy in view of his abilities is to transfer him or to change his job. This can be achieved by reorganisation and by the provision of job aids and back-up services. Where training is required for improvements in performance to be achieved the next decision is to choose the method. Training can be carried out 'on the job' or 'off the job'. On the job training is usually cheaper because it avoids the costs of lost time for work as well as accommodation, travel and other formal tuition costs. It is a suitable method in work situations which are simple, safe, stable, unlikely to change and do not have much time or cost pressure. 'Off the job' training will be necessary wherever:

(a) The job is complex, particularly if the complexity arises because of demands for information processing and decision making.
(b) The results of incompetence are large.
(c) There is constant change calling for rapid regrouping of knowledge and skills.
(d) The job forms part of a new operational system still under development.
(e) The job does not exist.
(f) There are motivational or attitudinal problems that cannot be tackled in a formal work environment.

Learning takes time. Initial improvements in skill are slowly acquired. Thereafter, a rapid improvement takes place. Refinements in skill are acquired after long experience. An essential element in learning is the confidence of the individual. Experience of success produces it and increases capabilities; failure does the opposite.

The problem for a training agency is to identify training needs and this is especially difficult in management. Managerial weaknesses are not easily measurable nor are solutions to inadequacies obvious. The conventional management course is normally about the application of techniques, for example in budgeting or investment appraisal, so that really hard everyday creative thinking required by management to solve their problems is missed. There are courses on methods of creative thinking but unless they are applied to real problems they can never be entirely satisfactory. Indeed there is a growing feeling that management tutors are equipped only to act as catalysts in management training. Their intellectual skills can assist in providing a rigorous approach to problems. The main thrust for training must be by the managers themselves as they define and solve problems together.[16] This is of course not a new concept rather a reaction to formal tuition. Practical skills have been passed on successfully within families and communities for generations without the benefit of outside agencies.

Training/education when done properly has a real payoff, not only in the way in which work is accomplished but also in the satisfaction achieved by the individual concerned. Work well executed is often a measure of status in an agricultural community and within organisations. This is partly because of most people's innate desire to do creative things but it is also true that it makes all the difference to performance. A good stockman in livestock enterprises has much higher performance levels than a poorer one, so much so that one of the keys to success in such enterprises is to acquire such men. In modern arable farming a single mistake can be extremely expensive. Task analysis or in other terms a job description is thus helpful in the selection and placing of individuals with the right capabilities within organisations.

COMMUNICATION

Communication is an important ingredient in the working of any part of an agricultural system. Information is passed from one person to another and nowhere is this better illustrated than in the market place. Price information is nowadays electronically distributed around the world, local newspapers print it and numerous informal contacts are made by telex, telephone and by word of mouth. Information on new technology and technological problems is conveyed back and forth between practitioner and researchers through extension workers and by other means. Whole libraries can be searched by means of abstracts and computerised searches. Accounts are prepared for farms, businesses and institutions and reports, instructions, directions, questions and answers are continually produced in the written and spoken word. Management is continually balancing the line between generating and distributing too little and too much information. Four basic questions need to be continually borne in mind. What is relevant, when is it needed and who gets it? The fourth question is how reliable does information need to be for it still to be used and not be misleading?

Information can be misleading because of what is not reported. For example a profit and loss statement, physical yields, cash flows or a balance sheet on their own can mislead and there is no way of checking on assumptions or wider issues. Similarly a reported fall in price may be explained by adverse weather conditions or simply a fall in quality.

Some information cannot be made available quickly enough, such as latest prices, census data or actual trading figures. Other information especially during negotiation and staff management may have to wait until the complete picture is available. It is bad management to raise expectations unjustly. Confidentiality is also extremely important in personal and financial matters. Gossip can destroy people and, when factions develop, whole organisations.

The presentation of information is also a matter of decision and will depend on a thorough analysis of the target and the type of information to be disseminated. An important element in this is cost. The more personal and detailed the information to be conveyed the higher the cost. But any evaluation of the cost of information should also take into account the benefits to be derived from using it. Indeed, the whole point of information is to produce better decisions so that the payoff is potentially large. This is especially so in the marketing of agricultural produce. Classification and description make it possible to reduce the costs of transactions when the necessity to inspect produce is removed. Transport and storage can also be simplified and its costs reduced if produce need not be kept separate. Signals about what the market requires can be much more specific if grades and other measures of produce quality are available.

The Capacity of Communication Channels

A point that is often overlooked in formal organisations is the capacity of communication channels. Organisation charts assume that information flows from top to bottom and sometimes in other ways without constraint and with little sideways movement. In fact there is continual contact, sometimes by chance, sometimes by design, between people at all levels in an organisation. The same is true for natural communities of animals and plants.[17]

Information passed down the line is subject to both distortion and disruption. Both increase, the more people information is passed through, due to imperfect transfer and delays. Disruption is also more likely if information has to pass through a single crowded channel. A simple example would be a system where all communication was by a single telephone. It is not hard to understand how eventually as the number of people wishing to speak to one person at the other end of the line increases, large queues develop. In fact they develop rather quickly and can become very long because *time cannot be stored* and messages of different lengths arrive irregularly.

If the listener has time to spare when no one is talking to him, this time cannot be saved until more than one person wishes to talk to him at the same time. Two quantities are important for this simple queuing example, namely, the time interval that elapses between different messages arriving, the inter-arrival time, and the time it takes to listen to the message, the service time. An unsuspecting manager might do the following calculation to determine the communication capacity of a single person. If the average length of a message is five minutes, then one person should be able to listen to 12 people per hour. In fact if this were tried the queue length would approach infinity, unless each of the 12 people telephoned at exactly prescribed five-minute intervals and their messages lasted exactly five minutes. What happens in reality is that both the inter-arrival times and message times have a skewed distribution. There are a lot of short inter-arrival times and messages and only a few long gaps between messages and not many long messages.[18,19]

The ratio of the average service time and the average inter-arrival interval is called the traffic intensity (t). A mathematical relationship shows that the average total waiting time including the time to give the message is equal to $1/1-t$ multiplied by the average service time. Thus, as t approaches unity the waiting time approaches infinity. Values of t of less than 0·7 have reasonable waiting times, but for only a slight increase beyond this value, waiting times increase very quickly indeed. This means that where traffic intensity is approaching 0·7 there is very little resilience in the system to cope with bursts of messages or a particularly long message or indeed to reduce queue length once it has built up. Management then in such situations has to consider rearrangements of the system. In our example, a rule such as that non-urgent messages should be sent by post would allow the listener to use his idle time reading. Flexibility could be induced by having spare listeners at busy times or by making sure that messages can be directed to those concerned without the need to pass through filters, censors or information officers.

The same type of capacity problem occurs in meetings where if protocol is observed all matters should be referred to the chairman, one person speaks at once and the rest listen or at least do not make a noise. The larger the committee, the greater the average time spent listening and committees thus make progress more and more slowly. Where creative or detailed work is required of a committee, it can only proceed if small groups are created, say less than 10 and ideally four or five people. Participative types of courses hit the same problem, the critical maximum size being around 15, the minimum size for interaction to occur around 10.

OVERCOMING BOTTLENECKS

Getting around bottlenecks can be achieved by changes in systems. Work study can reduce time delays by reorganising work after careful analysis although this cannot be done irrespective of cost. A good example of this is the effect of altering the shape and size of fields on work rates. Sturrock and Cathie[20] have shown that increases in the efficiency of work rate are of the order of 6–15% when moving from fields of 40 ha to 80 ha. Likewise the best shape of a field is a narrow rectangle. Buildings can be altered and designed to minimise the time and effort to do jobs. For example, a simple rule for jobs that require travel around a building with a single entrance is to organise the route so that the finishing point is at the same place as the starting point as shown in Fig. 8.1.

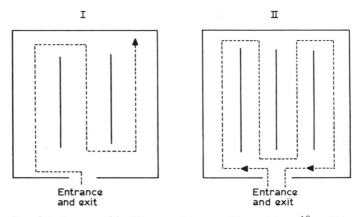

FIG. 8.1. Layout of buildings and routes. (From Moore[18] p. 67.)

Delays in agriculture can easily be magnified due to the sequential and seasonal nature of agricultural processes. The time of planting of crops has a large effect on yields; in general a critical period is available in which the ensuing yield will be relatively unaffected by the date, but outside this period, yields can be reduced at an exponential rate with time (Fig. 8.2).

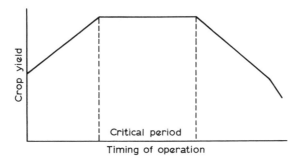

FIG. 8.2. Generalised effect of the timing of operations on crop yields.

A late sown crop will also tend to ripen and be harvested late which may then delay the preparations for the following crop. The costs of delay can dwarf cost savings through the better utilisation of machine capacity especially where these result in substantial falls in output. Capacity is not simply a physically determined concept but one determined by costs. Some delays may have to be tolerated since the costs of removing them are not justified.

Managing time in a personal sense is also about balancing the benefits of jobs completed against the costs of other jobs put off or not done. Lists of jobs to do are not a sufficient guide to action. Careful thought needs to be given to what is really important and the order in which jobs are done. Some jobs need to be done before others can start. Others can be done concurrently but they may compete for resources which are required for jobs with greater priority. Unnecessary tasks can be avoided if jobs are done in the proper order and care and foresight are taken. Three important concepts used in the technique of Critical Path Analysis[21] are helpful in determining priorities. Firstly, the earliest start time for a job. This depends on the completion times for jobs which allow it to begin. Secondly, the latest finish time for a job, that is the time beyond which a whole project is held up. Thirdly, the concept of the critical path, that is those jobs where the completion times are critical to the length of a total programme. Knowledge of the critical path is crucial in the setting of priorities.

ORGANISATIONS

Bottlenecks can also occur in agriculture because of organisational weaknesses. A systems or overall view is often lacking for problems such as conservation, pollution measures, erosion and flood control. The individual farmer or organisation can, as a small part of the whole, escape or impose costs and benefits on the rest of society. A simple case would be for drainage. A farmer as an individual may wish to drain his land. If he drains his land, he may increase the drainage problems of his neighbours lower down the valley. Alternatively he may not be able to carry out a drainage scheme because the cooperation of all the land owners in a poorly drained area is required. In similar ways water tables are lowered by independent operators of wells, irrigation water may be wasted near to its source but further down the distribution system supplies may be erratic and short. Shelter belts to protect valleys may not be planted, common pasture overgrazed, water courses and the atmosphere polluted and products not advertised. Unless some kind of public body is set up which by agreement or legislation brings *within the system* the costs and benefits of public responsibilities then all concerned are collectively worse off. It is rarely sufficient to leave public responsibilities to governments since unless local people are involved in managing their own destiny the perception that public bodies are run for their benefit can easily be lost.

CONCLUSIONS

Giving effect to plans and controls is dependent on the efficiency of implementation. Many personal skills are involved ranging from leadership, negotiation, organisation, and administration to communication. Implementation is not independent of design, since day-to-day operational snags can lead to new plans being proposed. But the best plans need skills such as time management, the selection, training and posting of staff and the ability to define, solve and prevent problems. This chapter has also tried to demonstrate that a single-minded approach to even highly desirable objectives can produce undesirable side effects. The less well-off can quite easily be asked to bear the greater part of the costs of change unless steps are taken to prevent this.

REFERENCES

1. HILL, P. (1970). *Studies in rural capitalism in West Africa*, African Studies Series. Cambridge University Press, Cambridge, Chapter 2.
2. SCHUMACHER, E. F. (1974). *Small is beautiful. A study of economics as if people mattered*, Abacus, Sphere Books, London (especially Chapter 16).

3. ARNON, I. (1968). *Organisation and administration of agricultural research*, Elsevier, Amsterdam, (especially Chapters 7 and 8).
4. HARRIS, T. A. (1970). *I'm OK–you're OK*, Pan Books Ltd., London.
5. BINSWANGER, H. P. (1978). *The economics of tractors in south east Asia: an analytical review*, Agricultural Development Council, New York and International Crops Research Institute for the Semi Arid Tropics, Hyderabad, India.
6. DILLON, J. L. (1976). The economics of systems research. *J. Agric. Syst.* 1, (especially pp. 11 and 12).
7. DONALDSON, G. F. and McINERNEY, J. P. (1973). Changing machinery technology and agricultural systems, *Am. J. Ag. Econ.*, 55, 829–839.
8. SCOBIE, G. H. (1979). *Investment in international agricultural research: some economic dimensions*, World Bank Staff Working Paper No. 361.
9. EVANSON, R. E. and KISLEV, Y. (1975). *Agricultural research and productivity*, Yale University Press, New Haven, Conn.
10. ARNDT, T. M., DALRYMPLE, D. G. and RUTTAN, V. W. (Eds.) (1977). *Resource allocation and productivity in national and international agricultural research*, University of Minnesota Press, Minneapolis.
11. DALTON, G. E. (1980). The educational role of farm management extension work by state advisory services, *J. Ag. Econ.*, 31, 2.
12. GALBRAITH, J. (1980). *American capitalism: the concept of countervailing power*, Boston, Mass, (2nd Ed.), Blackwell, Oxford.
13. MARTIN, A. (1958). *Economics and agriculture*, Routledge & Kegan Paul, London.
14. UPTON, M. (1976). *Agricultural production economics and resource use*, Oxford University Press, London, pp. 224–226.
15. DAVIES, I. K. (1971). *The management of learning*, McGraw-Hill, London.
16. REVANS, R. W. (1980). *Action learning. New techniques for management*, Blond and Briggs, London.
17. BOUGHEY, A. S. (1971). *Fundamental ecology*, Intertext Books, London.
18. MOORE, P. G. (1968). *Basic operational research*, Topics in Operational Research Series, Pitman, London, Chapter 5.
19. STAFFORD, L. W. T. (1978). *Business mathematics*, M & E Handbooks, London, Chapter 17.
20. STURROCK, F. G. and CATHIE, J. (1980). *Farm mechanisation and the countryside*, University of Cambridge, Department of Land Economy, Cambridge, Occasional Paper No. 12.
21. WIEST, J. D. and LEVY, F. K. (1969). *A management guide to PERT/CPM*, Prentice-Hall, Inc., Englewood Cliffs, New Jersey.

Index